A Guided Tour of the TI-85 Graphics Programmable Calculator

with Emphasis on Calculus

John F. Lucas
Mathematics Department
University of Wisconsin–Oshkosh

Christine A. Lucas
Mathematics Department
Whitefish Bay High School

Ardsley House, Publishers, Inc. • New York

Address orders and editorial
correspondence to:
Ardsley House, Publishers, Inc.
320 Central Park West
New York, NY 10025

ISBN: 1-880157-10-1

Printed in the United States of America

10 9 8 7 6 5 4 3 2

To Stephanie
and
to Our Students

Contents

Introduction

We wrote this booklet with a specific instrument, goal, and method in mind. The instrument is the Texas Instruments TI–85 graphics programmable calculator; the goal is to develop the user's skill so that he or she will feel comfortable employing the calculator as a valuable assistant within a variety of mathematical contexts; and the method is to guide the user through most of the commonly used menus and operations, with instant feedback by providing expected results and solutions to most problems. We have divided the booklet into nine individual ***Explorations***; there is also a reference section on Home screen or Programming formats as well as a set of menu and submenu displays.

This booklet can be used as a self-study guide or as a workshop manual. Students who are first learning the TI–85 could be expected to complete all Explorations in approximately 15–20 hours in an individual or group setting outside of class. Faculty learning the calculator in a workshop environment can study certain selected *Explorations* (or have these demonstrated) and reinforce their skills by completing the practice problems individually.

We believe this calculator and others like it will revolutionize the teaching and learning of classroom mathematics. For sheer power, it certainly does not replace a computer. But it wins hands down over the computer with respect to cost and portability for the student, and we believe these latter two considerations are now and will continue to be significant for the majority of students.

Finally, we selected the TI–85 for two reasons. We are both fulltime classroom teachers of mathematics. We had been working with the TI–81 in our respective classrooms at the university and high school levels over the past two years, and we are convinced that mathematics is "enlivened" through applications made more accessible by this technology. Secondly, both the TI–81 and TI–85 are menu-driven and quite user-friendly, so that the initial "learning time" is reasonable with respect to each. Currently, we do not see the TI–85 as a replacement for the TI–81. The more powerful TI–85 is perhaps best-suited to certain high school calculus courses and university mathematics programs involving calculus, matrix algebra, differential equations, and statistics. The TI–81 is still best-suited to precalculus, algebra, and other high school calculus courses. In a year or two our perceptions will probably change, as will the technology. For now, we are delighted to be in the midst of these exciting changes, and we hope you will be, too.

J. F. L.
C. A. L.

Some Information about the TI-85

- 32K RAM
- Stores/graphs up to

  99 functions
  99 polar equations
  99 parametric pairs
  9 first-order DE's
- Finds roots of up to 30th-degree polynomials.
- Can solve linear systems of up to 30 equations in 30 unknowns.
- Accepts unlimited number of matrices/vectors up to 30 × 30.
- Text screen size is 8 lines, 21 characters/line; menus appear in 7th and 8th lines.
- Graphics screen size is 126 pixels × 62 pixels.
- Computational accuracy is 14 digits (memory) with 3-digit exponent. (Screen display is 12 digits.)
- Uses 5 types of cursors:

  entry 

insert	_
2nd	↑
cap letters	A (boxed)
lower-case letters	a (boxed)

- Has seven regression (curve-fitting) models: linear, log, exp, power, and polynomials (2nd-, 3rd-, 4th-degree).
- Accepts nine types of data inputs for variables: numbers, vectors, matrices, lists, strings, equations, programs, graph databases, and graph pictures. Memory permitting, all variables can be named, stored, and recalled for later analysis.
- Uses 4 AAA alkaline batteries with a lithium cell (for battery-change backup).
- Is largely menu-driven.
- Has communications links to other TI–85s or a PC or Macintosh.

Exploration 1
Basic Moves, Menus, and Submenus

Press **ON**. Press **2nd MEM**, select **RESET**, select **ALL**, select **YES** . . . to clear the entire calculator. The top keys are *menu* keys. They are used to select menus that appear on the screen in the two lines directly above them. Press **GRAPH** to see one level of menus. Any of the F1–F5 keys select a menu. Press **MORE** to see five more menus; then **MORE** again to see the remaining three menus under the major heading "GRAPH." One more press of **MORE** returns you to the original set of menus.

Now select **ZOOM** and you now see two levels of menus. The top one is the old set that appeared in the 8th text line, now moved to the seventh text line. These menus are accessed by 2nd M1–M5 keys. The new menus appearing in the eighth line are submenus of the **ZOOM** menu. By pressing **MORE** several times, you can see there are 15 submenus inside **ZOOM** alone. You now should have a pretty good idea of the multiplicity and accessibility of menus in the TI–85.

To leave the menus, press **EXIT** once and the top menu disappears; then press **EXIT** again and the bottom menu disappears. (Once a menu is activated, pressing **CLEAR** removes both menus simultaneously.)

Let's graph a function and its derivative.

Press **GRAPH**, then **F1** (to select y(x) = menu), and press **SIN x-VAR ENTER** in response to y1 = . Then press **2nd CALC** to access the calculus menu, **F3** to select der1 (first derivative), and **2nd alpha Y 1** (this gives you a lower-case y), **,** (comma), and **x-VAR) ENTER**. Now press **EXIT** (to obtain 1st menu), select **RANGE (F2)**, and then set up the viewing window as you would on a TI–81. Press **ENTER** after each entry.

Use xMin = **⁻2π** (*Note:* a numerical approximation is inserted; also, use the *raised* negative key (⁻))

xMax = **2π**

xScl = **0**

yMin = **⁻1**

yMax = **1**

yScl = **0**

Then press **F5** (to graph). The graphs of the sine and cosine functions should appear, followed by the GRAPH menu in the eighth line.

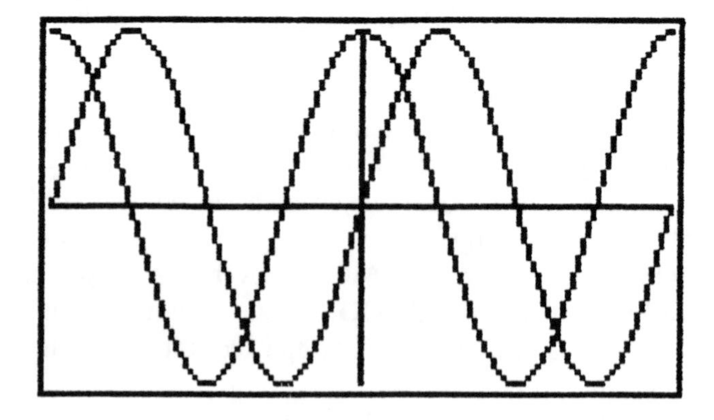

If you wish to remove the menu to see only the graphs, press **CLEAR**. If you wish to "hesitate" graphing—stop the graph and then start it from its last position—press **ENTER** to stop graphing, and then press **ENTER** again to start graphing again. Pressing **ON** also terminates graphing, but to continue, you must press **GRAPH** and the graph is redrawn from the beginning.

Note that the graphs of the sine and cosine were drawn *in sequence*, one after the other (y1 was drawn first, then y2). If you press **2nd MODE**, you will see a set of menus that are used to "set up" the calculator. The grey highlighted options are the default settings. You can move around these options using the cursor

left/right (◂ ▸)

or

up/down (▴ ▾)

keys. Once a setting is highlighted in grey winking format, you select it by pressing **ENTER**. The options shown here are *general* settings for the calculator. For example, suppose you cursor to **FLOAT**, then to 2, and select **2** by pressing **ENTER**. (This sets the numerical outputs to *two*-decimal-place accuracy.)

Next move down and select the setting **CylV ENTER**. This is the setting for vector (V) in cylindrical coordinates format (CylV). Return to the Home screen by pressing **2nd QUIT**. Then type in **2nd** [. (Use *square* brackets for vectors.) This is to be followed by **1, 2, 3 2nd**].

Now you see the 3-dimensional vector [1, 2, 3] on the Home screen, and if you press **ENTER**, you will obtain that vector converted to cylindrical coordinates:

[2.24 ∠ 1.11 3.00]

Note that the output does not contain commas. Also, the angle designator (∠) appears. So we have a vector with radial coordinate $r = 2.24$, angular coordinate $\theta = 1.11$ (radians), and z-coordinate, as before, at $z = 3$.

Press **2nd MODE** and change the decimal-place setting back to **FLOAT** and the coordinates display back to **RectV**. Then press **2nd QUIT** to return to the Home screen.

Now, let's return to our two graphs—press **GRAPH**, then **CLEAR** (to clean up the screen), and then press **EXIT** (once), then **MORE**, and select **FORMT** by pressing the **F3** key. This submenu, inside the GRAPH menu, is the graphing *format* menu. It looks like the 2nd MODE menu in its setup, but this particular menu relates to the graphing screen only. Suppose we select **DrawDot** (which creates dotted graphs), **SimulG** (which draws all selected functions simultaneously), and **AxesOff** (which removes the x- and y-axes from the graph window). Don't forget to press **ENTER** after you have cursored to each setting to select it. Now select **GRAPH** by pressing **F5**, and we see the effect of these different settings.

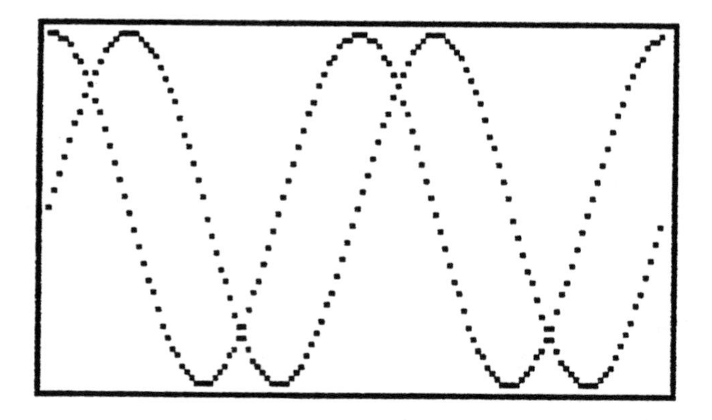

To see the entire graph, press **CLEAR** to remove the 8th line menus. Then, to return to the normal (default) graph settings, press **EXIT** (to recover the 8th line menus), **MORE** (to see five more menus), and **F3** (FORMT). Then cursor inside the graphing format menu to reset back to all original settings. Press **2nd QUIT** to return to the Home screen.

To clear graphs, enter the **GRAPH** menu, select **y(x)** = and deselect each function you don't wish to graph by placing the cursor on that function and pressing **F5** (the **SELCT** key). If you wish to deselect all functions at once, press **MORE** and select **ALL−(F2)**. Now press **2nd M5 (GRAPH)** to return to an empty viewing screen except for axes and the GRAPH menu. Press **EXIT** one or more times to leave a menu or submenu.

Practice Exercises

1. Graph the two functions

$$y1 = x^3 - 3x^2 + 1$$
$$y2 = x^2 - 6$$

in the viewing window (**RANGE**) $[^-2, 6]_x$ and $[^-7, 10]_y$ with scales set at 0.

Calculator Solution. Press **GRAPH F1** (y(x) = menu) and define these two functions (press **CLEAR** when the cursor is on each of the two original functions to "erase" them); then press **2nd M2** (**RANGE**) and define the given range settings; next press **F5** (**GRAPH**) to obtain the graphs in sequence. Once drawn, press **CLEAR** to remove the menus and see the entire graphs. You should see three distinct intersection points.

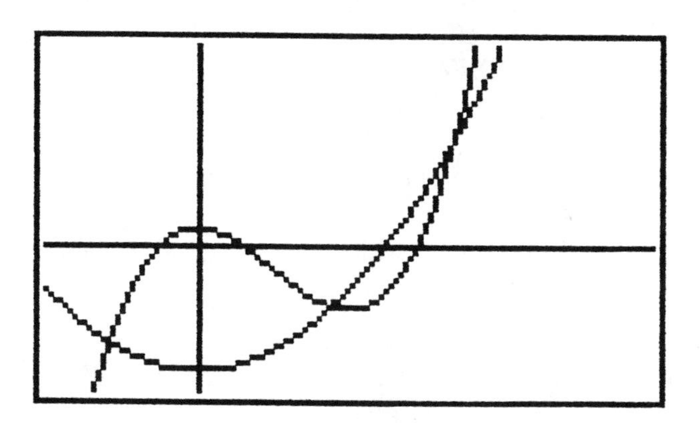

2. Now determine the coordinates of an intersection point of the two functions in Exercise 1.

Calculator Solution. Press **EXIT** to see the 8th line menus and select **TRACE** (by pressing **F4**). The cursor appears on the first selected function (here, y1). You can move it back and forth from one graph to another by using the up/down cursor (▲▼) keys. A little numeral appears in the upper right corner of the graphing screen to indicate the function on which the TRACE cursor is currently located. Trace along either curve (using ◂ ▸ keys) to a point near the intersection furthest to the right. Press **EXIT** to see the menus, and select **ZOOM** (by pressing **F3**). Now we see menus in both the 7th and 8th lines.

Those menus in the 8th line are ZOOM submenus; those in the 7th line are the original (GRAPH) menus that had been in the 8th line. Press **F1** to select **BOX**, and then use the cursor movement keys to move the cursor to the "northwest" corner of a boxing rectangle that will appear (to include that portion of a magnified graph showing the intersection more distinctly). Press **ENTER** to set the corner, then use the right cursor key ▸ and the down cursor key ▾ to "box" the region in. Press **ENTER** to set the "southeast" corner, and the portion of the screen inside the rectangle becomes the entire viewing window. (We can then **TRACE (2nd M4)** to determine a better approximation of the intersection point (which is about $x = 3.39138$, $y = 5.50147$), but there is a more efficient way described in Exploration 3.)

Finally, press **EXIT** to return to the menus; the ZOOM menu is now highlighted. Then press **F5 (ZPREV)** to return to the preceding viewing window. The two graphs will redraw, followed by menus appearing in the 7th and 8th lines.

3. From this point, exactly which key presses will enable you to redraw the graphs of these two functions *simultaneously*?

Answer. **EXIT MORE F3 (FORMT)**, cursor to select **SimulG ENTER F5** (for GRAPH).

Exploration 2

THE SOLVER MENU

Suppose we wish to solve an equation. For example, suppose we want to find the volume of a cylinder whose radius is 10 cm and whose height is 5 cm.

Press **2nd SOLVER** and carefully type in:

ALPHA ALPHA VOLUME = π **ALPHA (ALPHA ALPHA** RADIUS **ALPHA) x^2 (ALPHA ALPHA** HEIGHT **ALPHA) ENTER**

obtaining

eqn: VOLUME = π(RADIUS)^2(HEIGHT)

(Note the scrolling.)

Then enter values of 10 for RADIUS and 5 for HEIGHT, position the cursor on VOLUME, and either press **F5 (SOLVE)** or else type in a guess and then press **F5 (SOLVE)**. The output tells you VOLUME = 1570.7963267949, where the solution is between the lower and upper bounds shown (here corresponding to the real line), and the statement, left − rt = Ø, indicates that the equation is balanced.

We can solve for *any* variable in the equation in this way if the others are known. That is, if we wish to know the *radius* of a cylinder having height 5 cm and volume 1000 cm^3, first set the range (**F2**) at xMin = 0 and xMax = 500 to define a positive range for radius. (Since $(radius)^2$ appears in the equation, we can obtain either positive or negative values for it.) Then return to **SOLVER**, (press **2nd SOLVER ENTER**), set volume = 1000, **CLEAR** the entry for RADIUS, and select **SOLVE (F5)**. We obtain the result

RADIUS = 7.9788456080286

Type **EXIT** to leave the SOLVER. If you wish to store an unevaluated expression as an equation variable (for later analysis), follow these steps. Press **2nd QUIT** or **CLEAR** to produce the Home screen, then type out the equation (as you did previously inside SOLVER):

VOLUME = π $(RADIUS)^2(HEIGHT)$ **ENTER**

The calculator reports "Done." Then, to check that an equation variable has been stored, press **2nd VARS MORE**, and select **EQU (F3)**. You will then see

VARIABLES: EQU
▸ VOLUME EQU

Now press **EXIT** once, **2nd SOLVER**, and in response to eqn:, press **CLEAR** and select **VOLUM** from the menu, **ENTER**. You now see an equation exp = VOLUME, and you can solve for any of its variables as before by pressing **ENTER** and proceeding as before. Now, the variable for VOLUME is called "exp," which means an equation variable (actually, the right member of the equation, as an expression) has been stored under the name VOLUME. Press **EXIT** to leave the SOLVER menu and return to the Home screen. Press **CLEAR** to clear the Home screen.

Please note that throughout the exercises and examples in this book, your numerical results may vary slightly from those reported here. That is due to differences in our **2nd TOLER** menu settings. If the tolerance and delta values are set for a high degree of accuracy, we will tend to agree in more decimal places.

Practice Exercises

1. Solve the equation $e^x - 5 = 0$ for x.

   Answer. $x = \ln 5 \approx 1.609$

2. Solve the equation $e^x - 5x = 0$ for x.

   (a) Use a guess at x of 0.

   Answer. x = .25917110181906

   (b) Use a guess at x of 2.

   Answer. x = 2.5426413577736

   (c) Examine the situation graphically (inside **GRAPH**) by defining

   $$y1 = e \wedge x - 5x$$

   and setting the Range at ⁻2, 3, 0, ⁻5, 10, 0 and TRACEing along the curve near both roots.

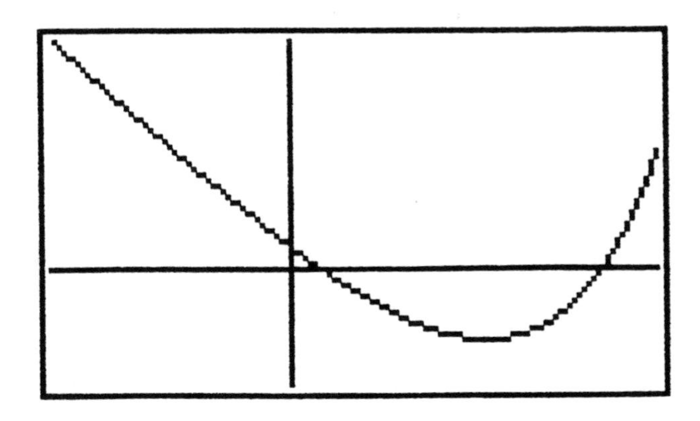

   (d) To get a better look at the smaller root, **TRACE (F4)** to a point near it, press **EXIT** (to see the menus), select **ZOOM**, then **ZIN ENTER**. The graph redraws. Position the cursor at the intersection by pressing **EXIT TRACE (2nd M4)**, and repeat the procedure until the desired accuracy is achieved. (Use EXIT each time to recall the menus.) Press **2nd QUIT** to start over.

3. Find the intersections of the functions

$$y = 2x^3 - 10x + 5 \quad \text{and} \quad y = x^2$$

Calculator Solution. Go to the **SOLVER** menu, **CLEAR** any equation that is there, define the equation $2x^3 - 10x + 5 = x^2$, and press **ENTER**. Set up a suitable range, like $[^-3, 3]_x$ and $[^-10, 20]_y$. **GRAPH (F1)** the function; what you see is

$$f(x) = 2x^3 - x^2 - 10x + 5$$

TRACE (F4) to a point nearby any root, press **EXIT**, and select **SOLVE**. Then select **GRAPH**, **TRACE** to another root, and repeat the procedure—**EXIT**, then **SOLVE**. Do the same for the third root.

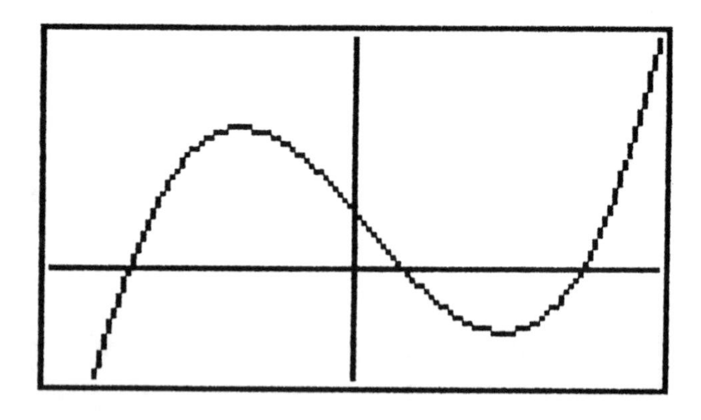

You should obtain

$$^-2.2360679774998, \quad .5, \quad 2.2360679774998$$

as the three roots. Press **2nd QUIT** to return to the Home screen.

4. The magnitude M of an earthquake on the Richter Scale is

$$M = 0.67 * \log_{10}(0.37E) + 1.46$$

where E is the energy of the earthquake in kilowatt-hours. Find the energy of an earthquake of magnitude 7. Using **SOLVER,** you should obtain 501712581.88983. Use the expressions MAG in your equation to indicate magnitude, and ENERGY for E.

When finished with these exercises, press **EXIT** or **2nd QUIT** to return to the Home screen.

Exploration 3 GRAPHING: MATHEMATICAL CHARACTERISTICS

For this example, we will use the pair of functions

$$y1 = 2x^3 - 10x + 5 \quad \text{and} \quad y2 = x^2$$

Select the **GRAPH** option, **y(x) =** , and define these two functions. If you desire sequential, rather than simultaneous, graphing, press **GRAPH MORE FORMT** and cursor to **SeqG**, selecting it by pressing **ENTER**. Set the range options (**F2** or **2nd M2**) at ⁻3, 3, 0, ⁻20, 20, 0, and **GRAPH** (**F5**) the functions. Now press **MORE** and select **MATH**.

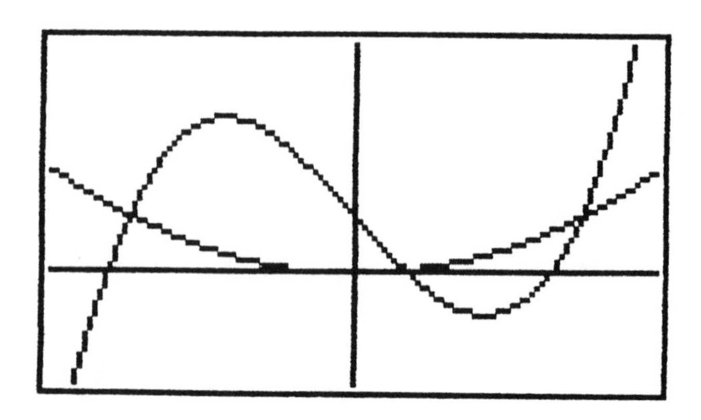

We will demonstrate all 13 submenus here.

LOWER/UPPER

Set a lower (or upper) x-bound as follows: select **LOWER (F1)**, then use cursor keys to position the cursor x-wise where you'd like a lower bound and press **ENTER**. A small marker ▸ appears above the graph. Next, select **UPPER (F2)**, then move the cursor to the right until you reach an upper bound for an interval and press **ENTER**. Another marker ◂ appears to indicate the upper bound. This does not reset the range, but it is used for finding roots, intervals of integration, and for other mathematical investigations. Reposition the lower and upper to $x = {}^{-}3$ and $x = {}^{-}2$, respectively. Also, press **GRAPH y(x) =** , place the cursor on y2 and deselect it (press **F5**).

ROOT

Now press **EXIT MORE**, and select **MATH (F1)** again. Select **ROOT (F3)** and move the cursor along the graph to the approximate position of the negative root. Then press **ENTER**. After a moment, we obtain

Root
x = ⁻2.453366647 y = 0

Press **EXIT** to return to the menus, then select **MATH (F1)**, and move the lower- and upper-bound indicators to bracket the next root, say about $x = 0$ and $x = 1$, respectively. Then press **ROOT (F3)**, position the cursor near the root, and press **ENTER**. The other two roots are .52972990065 and 1.9236367459. (Reposition the lower and upper bounds to ⁻3 and 3, respectively, when you're finished.) ROOT uses the SOLVER routine. Press **EXIT** to return to the GRAPH MATH menus, and select **MATH (F1)**.

dy/dx

When you select this option (**F4**) and move the cursor to a point on the graph, pressing **ENTER** reports the slope of the tangent (1st derivative) at that point. For example, the slope of the tangent at $x = 0$, $y = 5$ is $^{-}10$. Try several other locations. (Press **EXIT** to see menus, select **dy/dx**, move the cursor, and press **ENTER.**) Press **EXIT** to return to the GRAPH MATH menus.

∫f(x)

This approximates definite integration between any selected lower and upper bounds. For example, select **∫f(x) (F5)** and position the cursor left to $x = {}^{-}3$, press **ENTER**, then move the cursor right to $x = 3$, and press **ENTER**. Do not overshoot the left and right bounds; otherwise, you will have to reset the range. The calculator reports

$$\int f(x) = 30$$

What we have just calculated is the numerical value of the definite integral of $2x^3 - 10x + 5$ over the x-interval $[{}^{-}3, 3]$. (Press **EXIT** to return to the menus.)

FMIN

Since the minimum of the function in the viewing window is between 0 and 3, set the **LOWER** and **UPPER** bounds to these values first. Press **MORE** and select **FMIN (F1)**. Move the cursor near the minimum point (as a guess) and press **ENTER**. The calculator reports

FMIN
$x = 1.2909945463 \qquad y = {}^{-}3.606629658$

FMAX

This works the same way as **FMIN**. Press **EXIT** to return to the menus. Bracket the apparent maximum between preselected **LOWER** and **UPPER** bounds (press **MORE MORE**), select **FMAX** (by pressing **MORE**, then **F2**), move the cursor near the maximum point on the curve for a current estimate, and press **ENTER**. Here the FMAX is at

$$({}^{-}1.290994449,\ 13.606629658)$$

(Press **EXIT** to return to the menus.)

INFLC

It appears that an inflection point might be at or near x = 0, so bracket this by resetting the lower and upper bounds to include 0. Select **INFLC** (by pressing **MORE**, then **F3**) and position the cursor where you think it is. The cursor position is right at (0, 5), so move it off a little (to illustrate "guessing"). After some calculations the calculator indicates:

INFLC
x = 0 y = 5

(Press **EXIT** to return to the menus.)

YICPT

Selecting this option (**F4**) yields the y-intercept. Move the cursor away from the current location, (0, 5), and press **ENTER**. The calculator indicates

YICPT
x = 0 y = 5

ISECT

This finds the intersection points of *two* graphs, one at a time, assuming each such point is bracketed by lower and upper bounds. Return to the GRAPH menu by pressing **CLEAR** and **GRAPH**, selecting **y(x)** =, placing the cursor on y2, and pressing **F5** to select. Then press **2nd M5** (**GRAPH**) to graph both functions. Press **MORE**, select **MATH**, then reposition the **LOWER** and **UPPER** bounds to bracket the first positive intersection. Press **MORE**, select **ISECT**, move the cursor near the intersection inside the bounded region, and press **ENTER** (twice). The calculator confirms the two curves that are involved in the search for an intersection. Each time you press **ENTER**, observe the indicator numerals at the top right portion of the screen. The calculator reports

ISECT
x = .5 y = .25

Find the other two intersections. You should obtain

($^{-}$2.2360679775, 5) and (2.2360679775, 5)

(Don't forget to press **ENTER** *twice* after selecting **ISECT** and positioning the cursor near the intersection inside the bracketed region. The first **ENTER** identifies the second curve to intersect with; the second **ENTER** initiates the calculation.)

DIST

This calculates the straight-line distance between two points on the graph of one or more functions. First, reposition the lower and upper bounds at ¯3 and 3. Next, press **EXIT MORE MORE TRACE (F4)** and use the ▼ ▲ keys to place the cursor on graph 1. Now select **DIST** (press **EXIT MORE MATH MORE MORE**) and move the cursor to a point, say

x = ¯1.285714286 y = 13.606413994

Press **ENTER**, then move the cursor to another point (to the left or right of the first one), say x = 0, y = 5. As the cursor moves, it "drags along" a line segment anchored at the first point. Pressing **ENTER** yields

DIST = 8.7019206537

which is the *linear* distance between the two points. Press **EXIT** to return to the menus.

ARC

This operation measures the directed *arc length* along the graph of one function. Select **ARC**, press **ENTER** (cursor on (0, 5)). Move the cursor along the graph back to the point (¯1.285714286, 13.606413994), press **ENTER**, and you have

ARC = ¯8.757747783

Had the points been located in reverse order (left to right), the positive arc length would be returned. Press **EXIT** to return to the menus.

TANLN

This operation draws a tangent line to the graph of a differentiable function at a specific point. Select **TANLN**, move the cursor to any point (say x = 0), and press **ENTER.** The tangent line (here, the inflectional tangent) is drawn and its slope is reported as dy/dx = $^{-}10$. Press **2nd QUIT** to return to the Home screen; press **CLEAR** to clear it out.

Practice Exercises

1. Sketch a graph of the function

$$f(x) = x^5 - 5x^3$$

over the viewing window $[^{-}3, 3]_x$ and $[^{-}20, 20]_y$. First, deselect y1 and y2 and place this new function in y3. Sketch a graph.

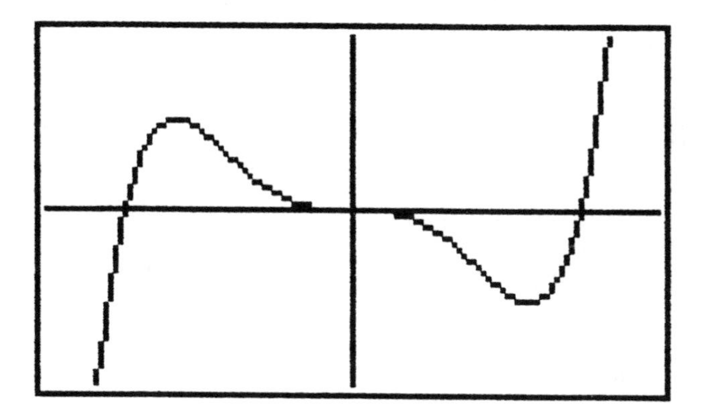

(a) Find each nonzero root.

Answer. The roots are given by $x = \pm 2.236067977$. Note that f is an *odd* function. Note also that, in exact form, the roots are $\pm\sqrt{5}$.

(b) Find the point where f has a local maximum.

Calculator Solution. Either return to the original range settings, select **FMAX** (in **MORE MATH MORE** and press **F2**), and position the cursor near the maximum, then press **ENTER**; or bracket the maximum via **LOWER** and **UPPER**, select **FMAX** (in **MORE** and press **F2**), position the cursor within the interval, and press **ENTER**. The latter procedure is better because it ensures a local extremum, rather than a global one. You should obtain

$$x = {}^{-}1.732050807 \qquad y = 10.392304845$$

Reposition the lower and upper bounds to $^{-}3$ and 3, respectively.

(c) Integrate f to approximate the area under the left "arch."

Calculator Solution. First select **∫f(x)**, (press **F5**), then position the integration bounds by moving the cursor along the curve to the root furthest to the left, pressing **ENTER**, moving the cursor to the origin, and again pressing **ENTER**). If the interval is $[{}^{-}2.236\ldots, 0]$, we obtain an approximate value of 10.4165637.

(d) There is an inflection point between the maximum point and the origin. Find its coordinates.

Answer. It should be

$$x = {}^{-}1.224744871 \qquad y = 6.4299105748.$$

(e) Approximate the slope of the inflectional tangent at the point of inflection you found in Part (d).

Calculator Solution. First, record precisely the x-coordinate of that point. Then press **EXIT MORE MORE** to select **dy/dx** (**F4**) or **TANLN** (**MORE MORE F3**). Next, cursor as close as you can to that value of x. At $x = {}^{-}1.238095238$, the slope of the tangent line is about $^{-}11.2$.

(f) Find the arc length of the complete graph over $[{}^{-}3, 3]_x$.

Calculator Solution. First, reset lower and upper bounds to $^{-}3$ and 3, respectively. Press **MORE MORE** and select **ARC** (**F2**). Then move the cursor left to the $x = {}^{-}3$ position (being careful not to overshoot the intended x-value), press **ENTER**, and move the cursor right to the x = 3 position. Press **ENTER** again and you should obtain (after waiting about 2¼ minutes!),

$$\text{ARC} = 258.36707677$$

If you wish to reduce the calculation time, consider a smaller interval, like $[0, 2]_x$. It still takes over half a minute, but the arc length here is 13.176422763.

(g) Finally, return to the **GRAPH** menu, select y1 (keeping y3 selected), graph both functions, and determine any intersection(s) that occur in the interval $[0, 2]_x$.

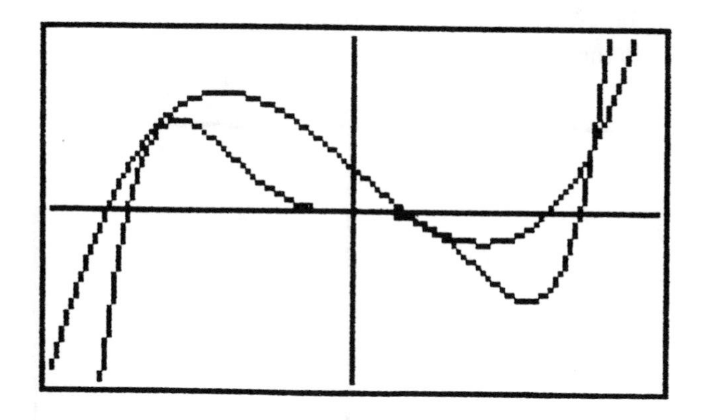

Calculator Solution. This is difficult to see, so select **ZOOM**, then **BOX**, and position the cursor first to the upper left vertex of the desired location of the box, **ENTER**, and move to the lower right vertex, and again, **ENTER**. What do you see? It appears there are *no* intersections there. Staying in the **ZOOM** menu, select **ZPREV**—this returns the window to its previous state. Let's search for the obvious intersection point to the right of x = 2. **EXIT** from the zoom menu (press **EXIT** *once*). Return to the **MATH** menu (press **MORE** and select **MATH (F1)**), bracket the intersection setting **LOWER** and **UPPER**, press **MORE**, select **ISECT**, cursor to the approximate intersection (along either graph and inside the bracketed region), press **ENTER** (twice), and obtain

ISECT
x = 2.364157097 y = 7.786106084

Press **CLEAR** (twice), **GRAPH**, select **y(x) =** , and **CLEAR** all functions previously defined.

2. Sketch a graph of the function

$$y = \left|\frac{\cos(25x)}{25}\right| + x$$

over the viewing range $[^{-}6, 6]_x$ and $[^{-}4, 4]_y$. Does this graph contain the origin?

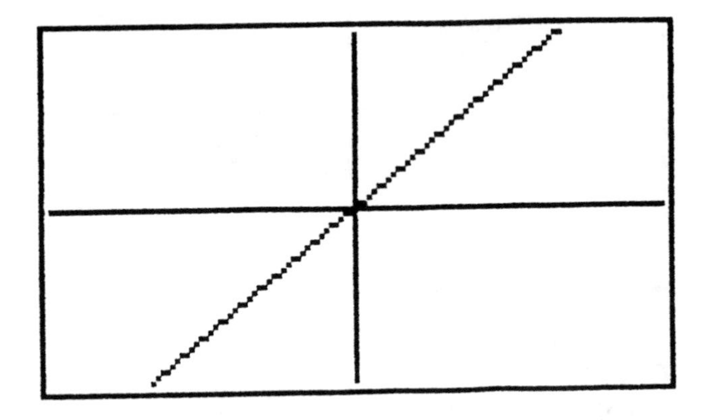

Calculator Solution. Note that absolute value is accessed by **2nd MATH**, select **NUM**, then **abs** (under **F5**). [If you use this function a lot, you can place it in the **CUSTOM** menu by copying it from the **CATALOG** as follows: **2nd CATALOG F3** (for custom) and **F1** to place it in the first position under the custom menu.] Next press **CLEAR GRAPH** and select **RANGE (F2)** to set as before. Once the function is graphed, it should look like the line y = x. Press **ZOOM ZIN ENTER**, and then **ENTER** one more time for a quick extra ZOOM to answer the question.

3. Try graphing and analyzing the function

$$g(x) = 2(x + 2)^{2/3} + \frac{(x - 4)}{(x^2 + 1)}$$

over the window $[^-3, 5]_x$ and $[^-2, 6]_y$.

Hint: Define the function by

$$2\left((x + 2)^2 \wedge \left(\frac{1}{3}\right)\right) + ((x - 4) / (x^2 + 1))$$

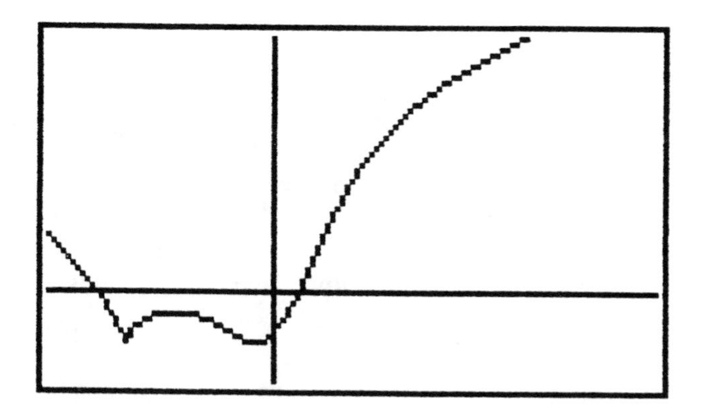

There is a singular point here (where the function is nondifferentiable). Don't attempt operations that involve derivatives near it. (You might try and see what happens.)

The remaining exercises do not include calculator directions, but provide answers you should expect if you use the calculator skillfully.

4. Consider the function $f(x) = \dfrac{x^3 - 10x^2 + x + 50}{x - 2}$.

   (a) Graph this function in the viewing window $[^-10, 10]_x$ and $[^-100, 100]_y$. Set the **GRAPH FORMT** to **DrawDot** option first to eliminate the vertical asymptote.

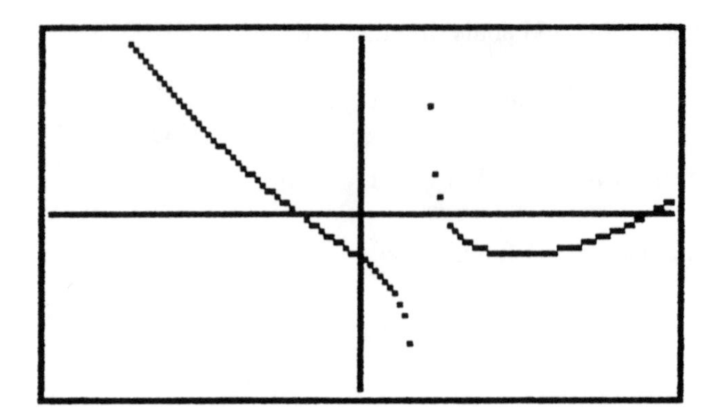

(b) Estimate all three zeros of this function.

Answer. $x = {}^{-}2$, $x = 2.6833752096$, $x = 9.3166247904$

(c) Estimate both coordinates of the minimum point on the right branch of the graph of the function f. (*Hint:* Bracket the region first by setting lower and upper bounds.)

Answer. $x = 5.0646965885$ $y = {}^{-}23.34048968$

(d) Long divide the numerator of f by its denominator. Sketch the parabola $y2 = x^2 - 8x - 15$ in the same viewing window as y1 given in Part (a). What is the relationship between the two graphs for very large x ($x \to +\infty$) or x large in absolute value, but negative ($x \to -\infty$)?

Answer. Since $f(x) = x^2 - 8x - 15 + \dfrac{20}{x - 2}$ and the remainder goes to 0 as $x \to \pm\infty$, then the parabola $x^2 - 8x - 15$ is a nonlinear asymptote for f. Reset the **GRAPH FORMT** back to **DrawLine** before proceeding.

5. Consider the function $g(x) = |x^3 - 15x + 1|$. (The **abs** function is located in **2nd MATH F1 (NUM) F5 (abs)**.)

(a) Sketch a graph of this function in the viewing window $[{}^{-}5, 10]_x$ and $[{}^{-}5, 30]_y$.

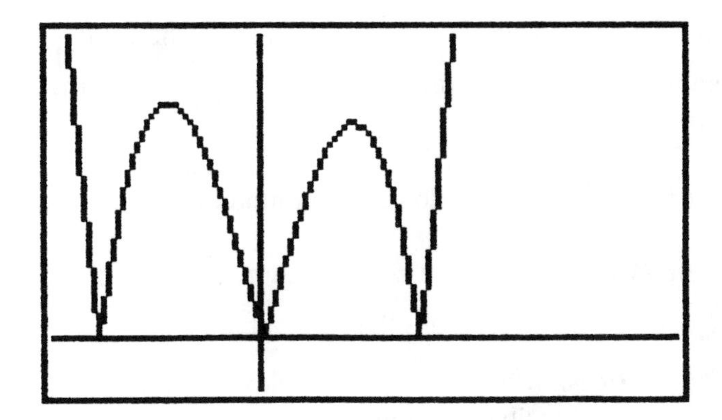

(b) Indicate the x-coordinates of the three minimum points (on the x-axis, so y = 0 in each) and indicate both coordinates of the two local maxima.

Answer. MINIMA at

$$x = {}^{-}3.905896302,$$
$$x = .06668673432,$$
$$x = 3.8392084188$$

MAXIMA at

$$x = {}^{-}2.236068705$$
$$y = 23.360679775$$

and at

$$x = 2.2360679468$$
$$y = 21.360679775$$

Note: The graph is *not* symmetric with respect to the y-axis.

6. *Application*: Nerve-Impulse Transmission

Nerve impulses in the body travel along fibers called axons. The axon transports the nerve impulse and has an insulated coating called a *myelin sheath.*

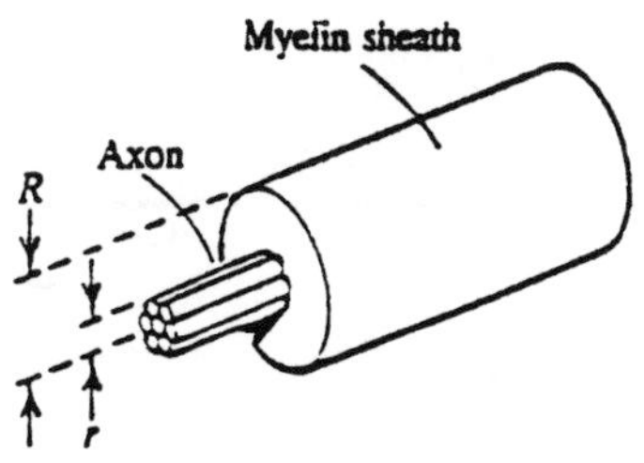

The velocity of an impulse v depends on the ratio $\frac{r}{R}$ of the radius r of the axon to the radius R (axon + sheath), by the formula

$$v(x) = -kx^2 \ln x$$

where $x = \frac{r}{R}$ and k is a constant of proportionality.

Basic Problem. Find the ratio $\frac{r}{R}$ that maximizes nerve impulse velocity v (in most humans this is about .6).

Calculator Solution. Set k = 1; graph $y = {}^{-}x^2 \ln x$ over $[0, 1]_x$ and $[0, .5]_y$. Obtain a maximum of

$$x = .60653332213 \qquad y = .18393972058$$

7. *Nonstandard Periodic Functions.*

   (Courtesy of Professor Edward H. H. Gade III, University of Wisconsin–Oshkosh Mathematics Department.)

The TI–85 can be used to demonstrate some rather interesting variations on nonstandard periodic functions using the greatest-integer function, **int** (located in **2nd MATH NUM**, select **F4**). In some of the situations given here, the **GRAPH FORMT** setting should be changed to **DrawDot**; in others, the **DrawLine** graphing format option returns a more pleasant picture.

(a) Sketch a graph of y1 = **int x** over $[{}^{-}4, 4]_x$ and $[{}^{-}4, 4]_y$ using **DrawDot** format. (Also, check what happens when you use **DrawLine** format.)

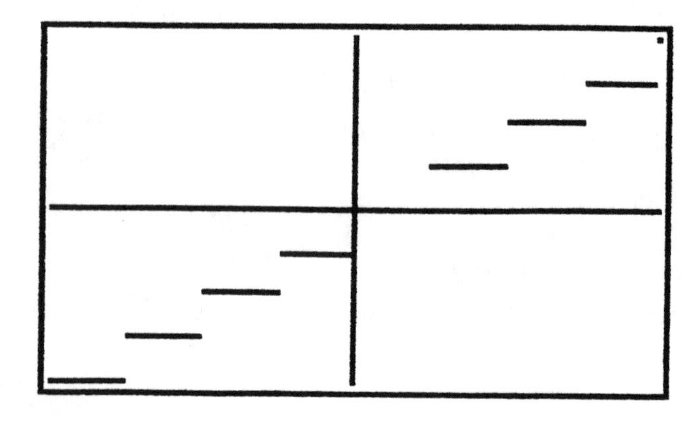

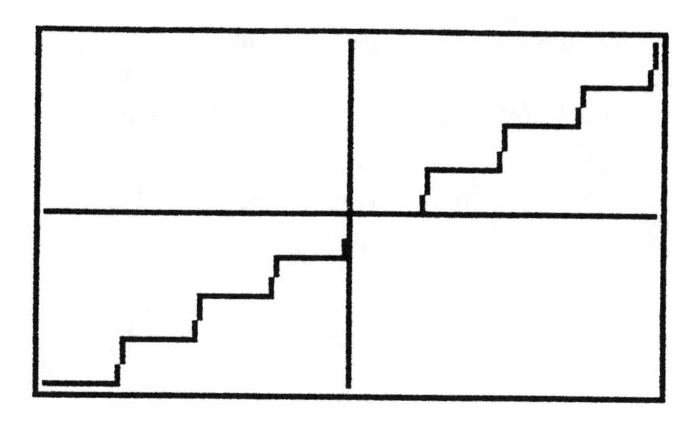

(b) Sketch the "sawtooth" function

$$y1 = \mathbf{x} - \mathbf{int\ x} \text{ over } [^{-}4, 4]_x \text{ and } [^{-}2, 2]_y.$$

using each format—**DrawDot**, then **DrawLine**.

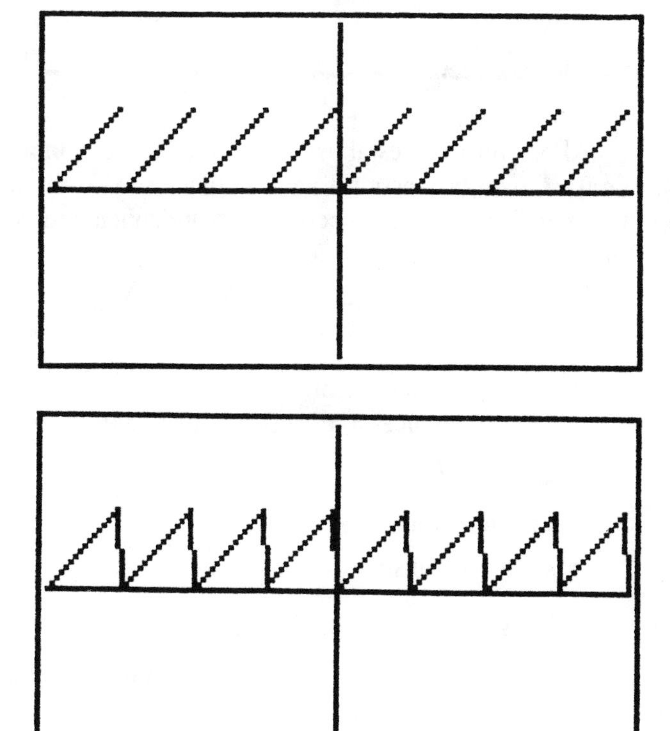

(c) If you have a function whose domain includes [0, 1), such as $f(x) = x^2$, you can obtain a periodic function (with period 1) by graphing its composition with the function x − int x. For example, try

$$y2 = (\mathbf{x} - \mathbf{int\ x})^2 \text{ over } [^{-}4, 4]_x \text{ and } [^{-}2, 2]_y$$

(You can deselect y1 and define $y2 = (y1)^2$.)

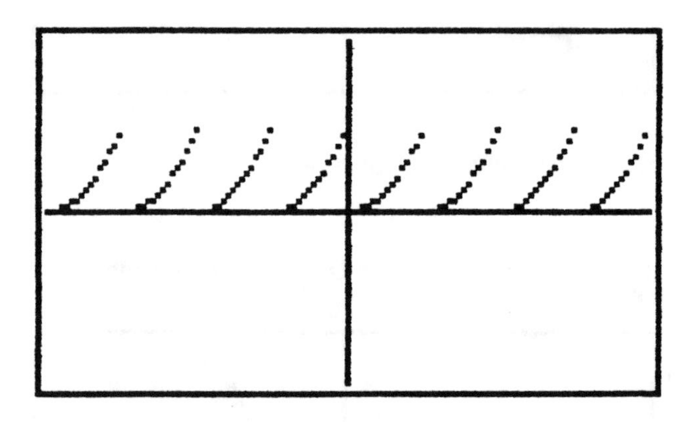

After you have established an interesting composition on [0, 1) with a function f and the function x − int x, then you can vary amplitude, period, and horizontal and vertical shifts by using the general form

$$y1 = a\,f(b(x - c) - \text{int}(b(x - c))) + d$$

where

a is the amplitude factor
(a < 0 produces a reflection in the x-axis)

$\frac{1}{b}$ is the period

c is the horizontal shift

d is the vertical shift

(d) If you were to double the period $\left(b = \frac{1}{2}\right)$, increase the amplitude by a factor of 3 (a = 3), and shift the graph of x − int x upward 1 unit, the appropriate function would be

$$y1 = 3\left(\left(\frac{x}{2}\right) - \text{int}\left(\frac{x}{2}\right)\right) + 1$$

Graph this variation over $[^{-}6, 6]_x$ and $[^{-}6, 6]_y$.

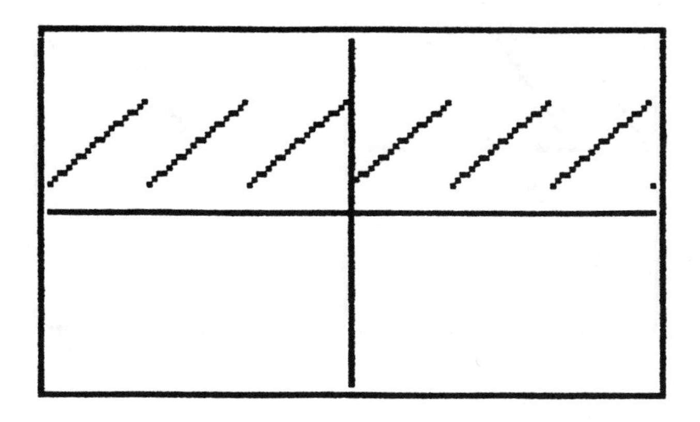

(e) Try the same variation as in Part (d), except with the $(x - \text{int } x)^2$ function. Here,

$$y1 = 3\left(\left(\frac{x}{2}\right) - \text{int}\left(\frac{x}{2}\right)\right)^2 + 1$$

Use the same window as in Part (d).

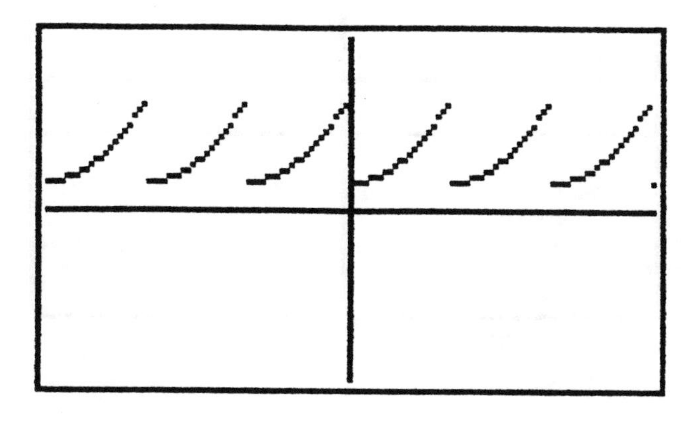

(f) Sketch a graph of

$$y1 = \mathbf{2x - 2 * int\ x} \text{ over } [^-4, 4]_x \text{ and } [^-3, 3]_y$$

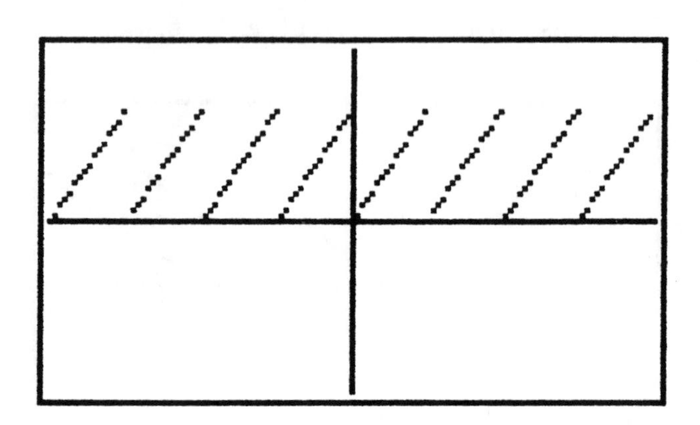

(g) Now try the following:

i. $y2 = \mathbf{int(2x - 2 * int\ x)}$ over $[^-4, 4]_x$ and $[^-3, 3]_y$.

Use y2 = int y1, using the y1 in Part (f). Deselect y1 first.

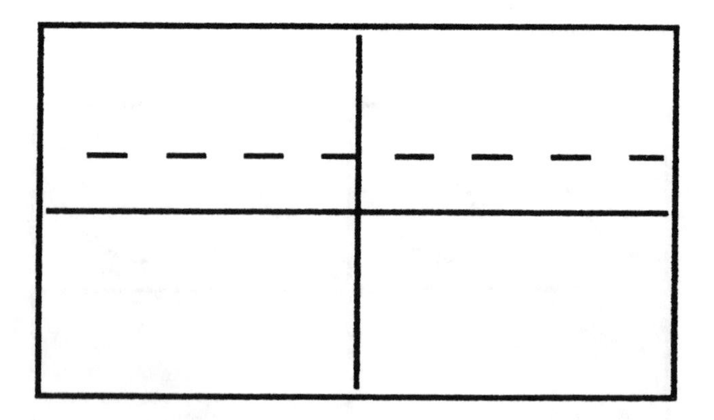

ii. $y3 = {}^{-}\mathbf{int(2x - 2 * int\ x)} + \frac{1}{2}$ over the same window.

Use $y3 = {}^{-}y2 + .5$. Deselect y2 first.

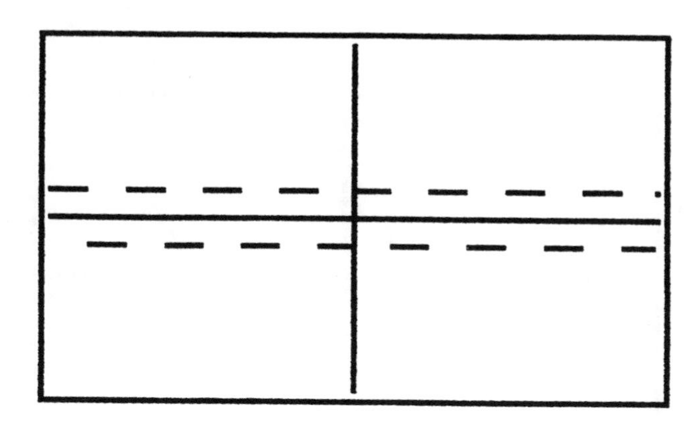

iii. $y4 = {}^{-}\mathbf{2 * int}\left(\mathbf{x - 2 *} \text{int}\left(\frac{\mathbf{x}}{\mathbf{2}}\right)\right) + \mathbf{1}$ (same window).

(*Note:* This is y3, with amplitude and period doubled.)

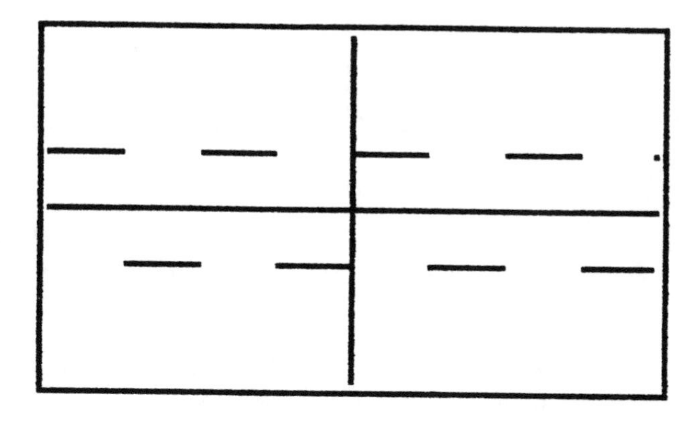

iv. This graph looks like sin(πx):

$$y1 = 4(x - \text{int } x - (x - \text{int } x)^2) * \left(1 - 2 * \text{int}\left(x - 2 * \text{int}\left(\frac{x}{2}\right)\right)\right)$$

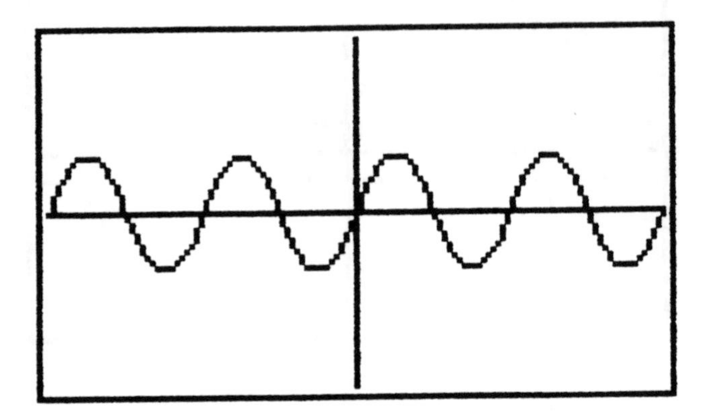

Now graph y2 = **sin(πx)** in the same viewing window.

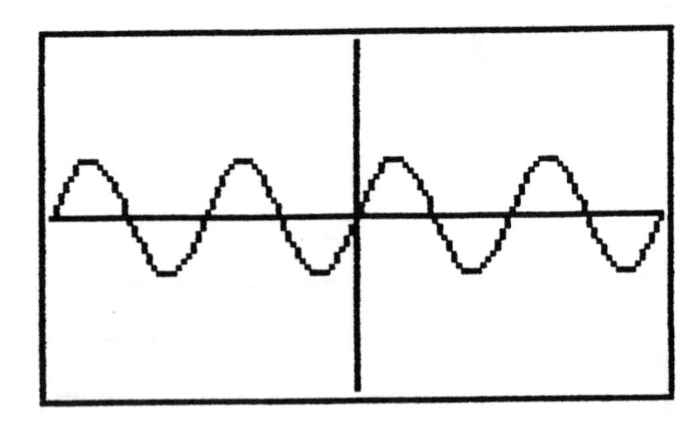

Don't forget to go to **GRAPH FORMT** *and change back to* **DrawLine** format *before proceeding to the next Exploration.*

Exploration 4 Numerical Calculus

From the Home screen, a variety of numerical calculus operations can be handled. Here are some examples of operations accessed by **2nd** CALC.

Evaluation of a Function: evalF

To evaluate a function at a point, we must select this option (**F1**) from the **2nd CALC** menu, define the function, the appropriate variable, and the point of evaluation.

Example:

To evaluate the function

$$f(x) = x^2 - 3x + 1 \quad \text{at } x = 2$$

we select **evalF (x^2 − 3x + 1, x, 2) ENTER**. The calculator indicates $^{-}1$ as the value of f(2).

Numerical Derivative: nDer

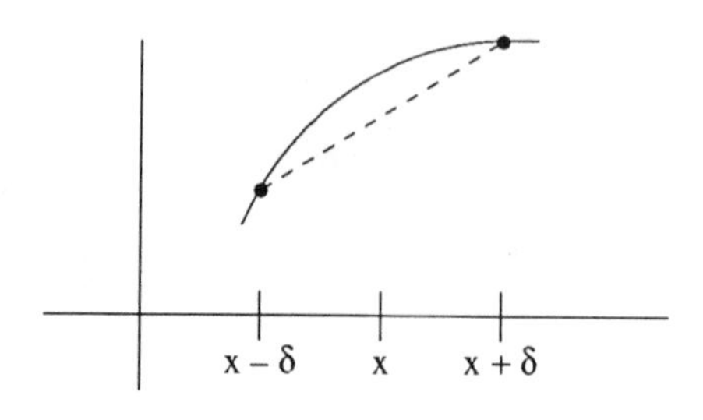

Using the variable δ in the TOLERANCE menu (with default .001), this operation approximates the slope of the tangent line to the graph of a function f at a value of a point x by computing the slope of the *secant* line, defined by a symmetric interval centered at x:

$$\frac{f(x+\delta)-f(x-\delta)}{(x+\delta)-(x-\delta)}$$

Format: **nDer(function, variable, value)**

Examples:

1. Approximate the slope of the tangent line to $y = x^3$ at $x = 2$. Also, give an exact value.

   Calculator Solution. **nDer (x ^ 3, x, 2)** yields 12.000001 **der1(x ^ 3, x, 2)** yields 12, because the first-derivative function is calculated *exactly* using the rules of differentiation.

2. You can nest **nDer** only once. A reasonable approximation to the 4th derivative of $y = x^5$ at $x = 2$ would be:

   nDer(nDer(der2(x ^ 5, x), x), x, 2)

   which returns 240. Be careful with parentheses here, and note that each successive derived function must have its variable defined, with the outermost being evaluated at 2.

First Derivative: der1 and Second Derivative: der2

Format: **der1 (function, variable, value); same for der2**

Examples:

1. Evaluate $D_x e^{^-x^2}$ at $x = -1$.

   Calculator Solution. **der1 (e ^ (⁻x²), x, ⁻1)** returns .735758882343.

2. Evaluate $D^2_{AB}(AB)^3$ at AB = 5.

   Note that here the variable is AB, and on the TI–85 juxtaposition does *not* imply multiplication for alphanumeric characters. Multiplication *is* implied by inserting a space between characters or by typing a * between characters.

   Calculator Solution. **der2 (AB ^ 3, AB, 5)** returns 30.

   How would you calculate a third derivative?

3. Approximate $f'''(1)$ if $f(x) = \tan^{-1}x$.

   Calculator Solution. **nDer(nDer(der1(tan⁻¹x, x), x), x, 1)** returns .499999.

Definite Integral: fnInt

Format: **fnInt (function, variable, lower limit, upper limit)**

Example:

$\int_1^3 x^2\,dx$ is computed by

fnInt (x², x, 1, 3)

which returns 8.66666666667.

(To convert this to a fraction, for purists, press **2nd MATH MISC MORE ▸Frac**, and you see

Ans ▸ Frac

Then press **ENTER**, and obtain 26/3.)

Return to numerical calculus by pressing **2nd CALC**. Then press **MORE** to see the remaining three operations.

Function Extrema and Arc Length: fMin, fMax, and arc

fMin, **fMax**, and **arc** yield each minimum location (x-wise), each maximum location (x-wise), and the arc length of a function, given its variable and lower and upper bounds. These bounds define approximate x-intervals on which these phenomena can be calculated.

Format: **fMin(function, variable, lower bound, upper bound); same for fMax and arc**

Examples:

Let's return to the function

$$y = 2x^3 - 10x + 5$$

which we saw graphically in Exploration 3. Now we'll do the same calculations numerically, but we must have some idea of where to look for the maximum and minimum values in order to establish brackets.

1. For the minimum location, which occurred at $x \approx 1.29$: Press 2nd **CALC MORE**

   fMin (2x ^ 3 − 10x + 5, x, 0, 3) ENTER

   returns 1.29099454626.

2. For the maximum location, which occurred at $x \approx {}^{-}1.29$: Type **2nd Entry** and edit the first statement to read

   fMax (2x ^ 3 − 10x + 5, x, ⁻3, 0) ENTER

   which returns ⁻1.29099454626.

3. Calculate the arc length of that piece of the graph from $x = {}^{-}1.286$ to $x = 0$. Press **2nd CALC MORE** and select **arc (F3)**.

   arc (2x ^ 3 − 10x + 5, x, ⁻1.286, 0)

   returns 8.75803439697.

Practice Exercises

1. Find the value of B that maximizes the function

$$A = \tan^{-1}\left(\frac{12}{B}\right) - \tan^{-1}\left(\frac{2}{B}\right)$$

over the B-interval [0, 10].

Calculator Solution. This kind of function occurs in the "best view" problem, where the height of a painting and the distance from its lower edge to eye level are given, and the problem is to find how far away (x) from the wall a viewer should stand to maximize the angle (θ) of view at his or her eye level.

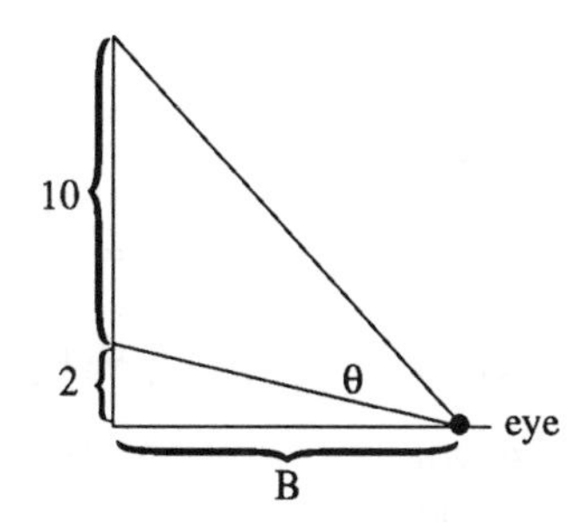

$$\textbf{fMax}\left(\tan^{-1}\left(\frac{12}{B}\right) - \tan^{-1}\left(\frac{2}{B}\right), B, 0, 10\right)$$

yields 4.8989782538. This would correspond to the number of feet a viewer should stand from a 10-foot painting hung 2 feet above eye level to achieve maximum viewing angle.

2. If $g(x) = (x^2 + 1)^{\ln x}$, find $g'(2)$.

Calculator Solution.

$$\textbf{der1}\left((x^2 + 1) \wedge \ln x, x, 2\right)$$

returns 4.14747885147.

3. Evaluate the following definite integrals, first numerically using the **2nd CALC** menu, then graphically, using the **GRAPH MATH** menu.

(a) $\int_{-2}^{5} \sqrt{x^3 - 2x + 7}\, dx$

Answer. 30.7963165704

(b) $\int_{0}^{\pi} x \cos x \, dx$

Answer. -2

(c) $\int_{-1}^{1} x\, |x|\, dx$

Answer. 0

4. Consider the two functions

$$f(x) = \sin\sqrt{(x - .5)} \qquad g(x) = \frac{(x^3 - 6x^2 + 5x + 2)}{8}$$

Determine the area of the plane region bounded by f and g.

Calculator Solution. In the **GRAPH** **y(x)** = menu, define y1 as f and y2 as g, and define y3 as y1 − y2. (**CLEAR** out any other functions that are there.)

Next, set the range values at 0, 6, 0, $^{-}2$, 4, 0, and sketch graphs of both functions (deselect y3).

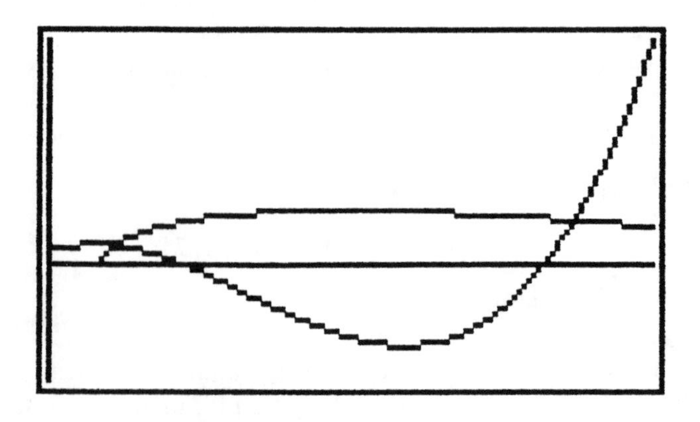

Now we can find their intersections in two ways:

1. Using the **GRAPH MATH** menu, setting bounds on each intersection, and using the **ISECT** operation, or
2. We could go to the **2nd SOLVER** menu and define the required equation as:

equ: y1 − y2 = ∅

Suppose we follow method 2. Press **ENTER**, then press **RANGE (F2)**, and press **ENTER** twice to select 0 and 6 as left and right bounds for x. Now, return to **2nd SOLVER** and **ENTER**. Then, using x = 1 as a guess, press **SOLVE (F5)** and obtain .64713744445829 as one intersection (x); still inside SOLVER, select **RANGE (F2)** and set xMin = **x** (use the x-VAR key, but this must be a lower-case x) and **ENTER** (the current value is that intersection location). Return to **2nd SOLVER ENTER**, use x = 5 as a guess, select **SOLVE (F5)**, and obtain 5.2099589301759 as the second intersection. Place this in **RANGE** as xMax (set xMax = **x ENTER**).

Finally, **EXIT** from SOLVER, go to the **GRAPH** menu, select **y(x) =** , deselect y1 and y2, and select y3; then graph that function over the present range.

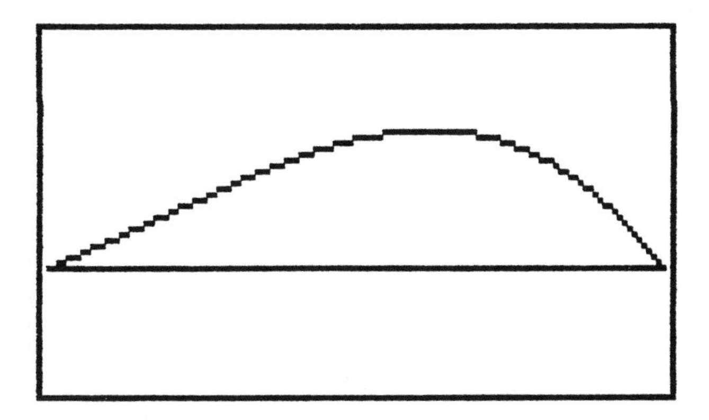

Now press **MORE**, select **MATH (F1)**, select **∫f(x) (F5)**, and move the cursor to the left bound of the window very carefully. (Don't overshoot the correct value, which is xMin, or .6471374444583). Press **ENTER**. Then move the cursor to the right bound of the window (xMax = 5.2099589301758) very carefully. (Don't go past the value.) Press **ENTER**, and the difference function is numerically integrated over the range determined by the x-intersections of the original two functions. We obtain:

$$\int_{.6471374444583}^{5.2099589301758} \left[\sin\sqrt{(x-5)} - \frac{(x^3 - 6x^2 + 5x + 2)}{8}\right] dx$$

$$\approx 6.8385718149$$

Press **2nd QUIT** and **CLEAR** to clear the Home screen.

Exploration 5 SYSTEMS OF LINEAR EQUATIONS

Consider the following system of linear equations:

$$\begin{aligned} 2x + 2y - 4z &= 17 \\ -x + y + z &= 0 \\ 5x + 6y + 2z &= 28 \end{aligned}$$

Solution by SIMULT Menu

The TI–85 will handle solvable systems from 2 to 30 linear equations simultaneously. Press **2nd SIMULT**, respond **3**, **ENTER**, to Number= (number of equations). You will then see a1, 1x1 . . . a1, 3x3 = b1 with a list of coefficients and constant b1 for the first equation. Enter these numbers, pressing **ENTER** each time. After b1 is entered, the next page requests the same information for the second equation, and so on. When the last set is entered, press **F5** (**SOLVE**), and the solution

$$\begin{aligned} x1 &= 2 \\ x2 &= 3.5 \\ x3 &= {}^{-}1.5 \end{aligned}$$

appears. On the TI–85, this is the fastest way of solving a system of linear equations.

Solution by MATRX Menu

The TI–85 will handle matrices up to size 255 × 255. We represent the preceding system as

$$\begin{bmatrix} 2 & 2 & ^{-}4 \\ ^{-}1 & 1 & 1 \\ 5 & 6 & 2 \end{bmatrix} \begin{bmatrix} x \\ y \\ z \end{bmatrix} = \begin{bmatrix} 17 \\ 0 \\ 28 \end{bmatrix}$$

or, in condensed form,

$$AX = B$$

Press **2nd MATRX** and select **EDIT** from its menu (key **F2**). Name the first matrix **A** and press **ENTER**. Next, change the dimensions to 3 × 3, in response to the position of the winking cursor, and press **ENTER** each time.

To input the matrix entries, first *note* that they are in *column* form, not row form. (This is a *difference* from the TI–81, where the entry format for matrices is row by row.) If you wish row-by-row entry, press **ENTER** after each matrix entry. Here, we will use column-by-column entry, pressing ▼ (down cursor) after each matrix entry.

So respond:

1, 1 = **2** (Press ▼ key, *not* ENTER for each entry.)
2, 1 = **⁻1**
3, 1 = **5**

and so on. Be careful to keep track of which entry you are making each time—when each page is flipped, the old user-entries remain and it can be confusing. When you are finished, press **2nd MATRX**, and select **EDIT** (**F2**) and input the B matrix, which is 3 × 1, with entries 17, 0, 28. If you press **2nd MATRX** and select **NAMES**, you will see the names of your two matrices A and B. They are also stored as matrix variables in the **2nd VARS MORE MATRX** menu. Now leave the **MATRX** menu by pressing **EXIT** twice, and have a look at your matrices on the Home screen by depressing **ALPHA A ENTER**.

Note: These matrices could have been constructed directly from the Home screen without **MATRX** by entering them using brackets, as follows:

A = [[2, 2, ⁻4] [⁻1, 1, 1] [5, 6, 2]]

A would then be stored as a matrix variable (whose entries are vectors), and it would appear under a **NAME** in the **MATRX** menu.

Since the solution matrix X is $A^{-1}B$, press

ALPHA A 2nd x^{-1} ALPHA B ENTER

and you obtain

```
[ [ 2   ]
  [ 3.5 ]
  [⁻1.5] ]
```

which is the solution of the system, expressed as a column matrix. This latter matrix will not be stored unless it is named as a variable.

Solution by Other MATRIX Operations

Once matrix A and matrix B are defined (and stored), return to **2nd MATRIX**, select **OPS (F4)** (for operations), press **MORE**, and select **aug (F1)** (for augmenting 2 matrices). Then enter

ALPHA A , ALPHA B) ENTER

and you obtain the matrix A augmented by B.

```
[ [  2  2  ⁻4  17]
  [ ⁻1  1   1   Ø]
  [  5  6   2  28] ]
```

Press **STO ▸ C** (stores as matrix C) **ENTER**; press **MORE** twice and select **rref (F5)** (for reduced row-echelon form), then **ALPHA C ENTER.** You now have the solution of the original system appearing in column 4 and the 3 × 3 identity matrix to the left of that.

```
[[1 Ø Ø  2  ]
 [Ø 1 Ø  3.5]
 [Ø Ø 1 ¯1.5]]
```

Determinant Equation of a Line

The equation of a line in a plane has the form

$$ax + by + c = 0$$

Now, if two points (x_0, y_0) and (x_1, y_1) are on this line, we could write a system of 3 equations in 3 unknowns, a, b, and c.

$$\begin{cases} xa + yb + 1c = 0 \\ x_0a + y_0b + 1c = 0 \\ x_1a + y_1b + 1c = 0 \end{cases}$$

This system is represented by

$$\begin{bmatrix} x & y & 1 \\ x_0 & y_0 & 1 \\ x_1 & y_1 & 1 \end{bmatrix} \begin{bmatrix} a \\ b \\ c \end{bmatrix} = \begin{bmatrix} 0 \\ 0 \\ 0 \end{bmatrix}$$

and it has a nontrivial solution if and only if

$$\det \begin{bmatrix} x & y & 1 \\ x_0 & y_0 & 1 \\ x_1 & y_1 & 1 \end{bmatrix} = 0$$

An Application to Geometry: We can use the determinant form of a linear equation to test to see if two given points (p, q) and (r, s) lie on *the same* side or on *opposite* sides of the line determined by points (x_0, y_0) and (x_1, y_1). Compute two determinants:

$$\det \begin{bmatrix} p & q & 1 \\ x_0 & y_0 & 1 \\ x_1 & y_1 & 1 \end{bmatrix} \quad \text{and} \quad \det \begin{bmatrix} r & s & 1 \\ x_0 & y_0 & 1 \\ x_1 & y_1 & 1 \end{bmatrix}$$

If these two determinants have the *same* sign, the points lie in the same half-plane; if they have *opposite* signs, the points are on opposite sides of the line.

Practice Exercises

1. Solve the following 4 × 4 system of equations by each of the three methods demonstrated here. Name your matrices D, E, and F (augmented).

$$\begin{aligned} {}^{-}x + y + 2z - 2w &= {}^{-}1 \\ 3x - 3y - 6z + 12w &= 3 \\ {}^{-}2x + 4y + 3z + 2w &= 3 \\ {}^{-}5x + y + 17z - 24w &= {}^{-}20 \end{aligned}$$

Answer. $x = 1,\ y = 2,\ z = {}^{-}1,\ w = 0.$

2. Consider the two 2 × 2 matrices

$$G = \begin{bmatrix} 1 & {}^{-}1 \\ 2 & 3 \end{bmatrix} \quad \text{and} \quad H = \begin{bmatrix} 0 & 4 \\ {}^{-}5 & {}^{-}2 \end{bmatrix}$$

Enter these, using the **MATRX** editor, and explore the **MATRX** submenus and keyboard to perform the following operations: (First, **EXIT** (twice) to leave the **MATRX** editor and go to the Home screen, then return for operations and functions.)

(a) det (G + H)

Calculator Solution. (Press **2nd MATRX MATH** to select **det.**) The answer is 10.

(b) G^3

Calculator Solution. Type **ALPHA G ^ 3 ENTER**. The calculator displays:

```
[[⁻9 ⁻11]
 [22 13 ]]
```

(c) $(G*H)^T$

Calculator Solution. This calls for the transpose of the product of G with H. Type (G * H) and select T by pressing **F2** in the MATRX MATH menu. (Use * or space to indicate multiplication.) The calculator displays:

[[5 ⁻15]
[6 2]]

(d) $H^T \ G^T$

Calculator Solution. See Part (c).

(e) G^{-1}

Calculator Solution. (Use **2nd x^{-1}** key.) The calculator displays:

[[.6 .2]
[−.4 .2]]

(f) $(5G + (.2H))^{-1}$

Calculator Solution. The calculator displays:

[[.131768953069 .037906137184]
[⁻.081227436823 .045126353791]]

(g) "Swap" rows 1 and 2 of G.

Calculator Solution. (Use **rSwap (G, 1, 2)** in **MATRX OPS MORE** MENU.) The calculator displays:

[[2 3]
[1 ⁻1]]

(h) Add row 1 to row 2 and store in row 2 of H.

Calculator Solution. Use **rAdd(H,1,2)**. The calculator displays:

[[Ø 4]
[⁻5 2]]

(i) Multiply row 2 of G by .5 and add to row 1.

Calculator Solution. Use **mRAdd(.5, G, 2, 1)**. The calculator displays:

[[2 .5]
[2 3]]

3. Consider the matrix

$$J = \begin{bmatrix} 1 & 1 & 2 \\ 0 & 1 & 0 \\ 0 & 1 & 3 \end{bmatrix}$$

Find the eigenvalues and the corresponding eigenvectors of J.

Calculator Solution. This is in the **2nd MATRX MATH** menu, keys **F4** and **F5**. We obtain

eigVl J ENTER

and the calculator display is

{1, 3, 1}

For **eigVc J ENTER**, the calculator display is

```
[[1 1 Ø   ]
 [Ø Ø 1   ]
 [Ø 1 ⁻.5]]
```

The columns are the eigenvectors corresponding to each eigenvalue.

4. Suppose airplanes landing at an airport (location (0, 0)) must always pass through the two points (100, 10) and (150, 12). A building is to be constructed at location 200. Can it be 15 units high? (Assume one unit = 10 ft.)

Partial Calculator Solution. Set up the determinant form of the line containing the two given points, and test (200, 0) and (200, 15) to see what their relation to the line is.

Exploration 6

SERIES AND SEQUENCES

On the TI–85, we can give a sequence of commands, each separated by a colon (above the • key), without having to write a program. Here, we examine how to work with series approximations, using the **seq** and **sum** keys of the **LIST OPS** Menu or **MATH MISC** Menu.

Compute

$$\sum_{n=1}^{100} \left(\frac{1}{2}\right)^n$$

First, we establish some variables. Let A be the first term, U be the upper index of summation, N be the variable of summation, and let L be a list (set) variable that stores all the terms from N = 1 to N = U. The **seq** and **sum** operations noted are located in **2nd LIST OPS MORE** (or **2nd MATH MISC**), keys **F3** and **F1**, respectively.

Now enter from the Home screen. (Before you do this, see the note following the entry sequence.)

100 → U: 1/2 → A: seq(A ^ N, N, 1, U, 1) → L: sum L ENTER

Note: By using A as the name of a *real-number variable* here, the *matrix* that was stored in A (Exploration 5) is lost. You may wish to use a different capital letter than A in the entries on pages 46 and 48. The → command is obtained by pressing **STO▸**. Once this key is pressed there is no need to press **ALPHA** before the variable name.

What happened? The sequence command (**seq**) created a list of values of the terms of $\left(\frac{1}{2}\right)^N$ with variable N, beginning at 1, ending at U, with increment 1. This list is stored as a list variable named L. To see this list, type **2nd VARS**, select **LIST (F4)**, then select **L**, and press **ENTER**. You are now looking at the first 100 terms of this geometric series (with appropriate scrolling).

If you wish to see all the elements of the list converted to exact (fraction) form, press

ALPHA L 2nd MATH MISC (F5) MORE ▸ Frac (F1) ENTER

The calulator displays

L ▸Frac
{1/2 1/4 1/8 1/16 1/...

Then, the **sum** command acts on lists and adds all members therein. We obtain a sum of 1, which is the limit to which this infinite geometric series converges. Apparently, the sum of the first 100 terms approximates the limit quite well.

A general format of commands for approximating any series is:

upper index → U: **seq**(term, variable, lower index, upper index, increment) → name for a list variable: **sum** list variable name.

Note that the seq command has five arguments and produces a *list* (or set). The sum command acts on that list, adding all elements, and producing an outcome that is a *number*. The general format can also be shortened by just using the seq command and typing in the specific arguments for the situation—then storing the result as a list variable and finally summing the list, as follows:

seq ((1/2) ^ N, N, 1, 100, 1) → S: sum S ENTER

Practice Exercises

1. Return to the **2nd LIST OPS MORE** menu and rewrite the set of commands to approximate

$$\sum_{n=1}^{100} \left(\frac{1}{3}\right)^n$$

Calculator Solution. Don't examine the list variable to see what entries are in the list here, because if no instructions are intervening, you can use the **2nd ENTRY** key each time to reproduce the commands and edit for new series. Use **INS** and **DEL** appropriately to edit. You obtain

100 → U: 1/3 → A: seq(A ^ N, N, 1, U, 1) → M: sum M ENTER (returns .5 after about 20 seconds)

The same result could be obtained by entering:

seq ((1/3) ^ N, N, 1, 100, 1) → M: sum M ENTER

Now, press **2nd ENTRY** to return to the command lines and edit them each time to do Exercises 2–5.

2. Compute

$$\sum_{n=1}^{50} \left[\frac{3^{n+1}}{5^{n-1}}\right]$$

Calculator Solution. This is really geometric, and can be rewritten as

$$\sum_{n=1}^{50} 9\left(\frac{3}{5}\right)^{n-1}$$

seq (9(3/5) ^ (N − 1), N, 1, 50, 1) → M: sum M ENTER

Alternately, you could enter the sequence term directly as it appears after **seq**. That is,

seq((3 ^ (N + 1))/(5 ^ (N − 1)), N, 1, 50, 1) → L : sum L ENTER

The answer is given as 22.4999999998. (The series converges to 22.5.)

3. Compute

$$\sum_{n=2}^{50} \left[\frac{1}{n(\ln n)^2} \right]$$

Answer. 1.85477071837

4. Approximate the sum of the convergent alternating series

$$\sum_{n=1}^{\infty} (-1)^{n+1} \frac{2^n}{n!}$$

by summing the first 10 terms.

Calculator Solution. The factorial symbol (!) is located in the **2nd MATH PROB (F2)** menu (option **F1**). The answer is .864620811287.

5. Compute the sum of the first 1000 terms of the (divergent) harmonic series

$$\sum_{n=1}^{1000} \frac{1}{n}$$

Answer. 7.48547086055

Note: This uses *lots* of memory. Check **2nd MEM** and select **RAM (F1)**; then examine the **LIST** memory used. You might wish to do a "garbage dump" by selecting **DELET (F2)**, pressing **F4 (LIST)**, and placing the cursor on L (assuming L was your last list variable), then **ENTER**. Following this, press **2nd QUIT** and **2nd ENTRY** to return your set of commands to be modified for this problem.

6. List the first seven terms of the sequence

$$\left\{ \frac{4n-3}{2^n} \right\}$$

in fraction form.

Calculator Solution. ▸ **Frac** is located in **2nd MATH MISC MORE**, option **F1**.

seq((4N − 3)/(2 ^ N) ▸ Frac, N, 1, 7, 1) ENTER

The output is

{1/2 5/4 9/8 13/16 1...

Note: The ▸ **Frac** command could have been placed after the entire sequence comand, with an identical output.

7. State the 4th partial sum, in fraction form, of the harmonic series

$$\sum_{n=1}^{\infty} \frac{1}{n}$$

Calculator Solution. Use **2nd ENTRY** to adjust your last set of commands to fit this. The sum is easily hand-calculated as $\frac{25}{12}$.

seq(1/N, N, 1, 4, 1) → M: sum M ▸ Frac ENTER

The answer is $\frac{25}{12}$.

8. Approximate $\int_0^1 2^t\,dt$ by lower and upper Riemann sums, where n (the number of subdivisions) is 100.

Calculator Solution. Since n = 100 and $\Delta x = .01$, we enter:

(a) For the lower sum:

seq((2 ^ t) * (0.01), t, 0, 0.99, 0.01) → S: sum S ENTER

The calculator indicates: 1.43770081711

(b) For the upper sum:

seq((2 ^ t) * (0.01), t, 0.01, 1, 0.01) → S: sum S ENTER

The calculator indicates: 1.44770081711

Note: By using the **2nd ENTRY** key to reproduce the set of similar commands each time, we can easily adjust to compute lower or upper sums for varying numbers of subdivisions, thereby obtaining a sense of the nature of convergence.

Exploration 7 Descriptive Statistics and Curve-Fitting

Descriptive Statistics for a Set of Test Scores

Suppose you have a data set of test scores, as follows:

Score	Frequency
98	3
95	2
92	3
89	5
85	4
82	1
79	3
75	10
70	5
65	8
60	1

Call up the **STAT EDIT** menu, and type in SCORE in response to the xlist name, press **ENTER**, and type in FREQ in response to the ylist name. (These names will be stored as list variables and can be recalled by name from the Home screen or by selection in the **STAT EDIT** menu.) Now enter the data in response to the item list, placing scores as x's, frequencies as y's.

When finished, **EXIT** once, select **CALC (F1) ENTER** twice (or use the down-arrow cursor ▾ twice). You should see the list names separated horizontally at the top. Now press **1-VAR (F1)**, and the calculator reports the descriptive statistics for the distribution:

$$\begin{aligned} \bar{x} &= 78.7555555556 \\ \Sigma x &= 3544 \\ \Sigma x^2 &= 284356 \\ Sx &= 10.9194479455 \\ \sigma x &= 10.7974391111 \\ n &= 45 \end{aligned}$$

If you wish to see a histogram of this data, **EXIT** twice, go to the **GRAPH** menu, deselect all functions in the y(x) = list, and set a reasonable range for this data, say 50, 100, 5, 0, 15, 1. Then press **STAT**, select **DRAW (F3)**, then **HIST (F1)**, and **CLEAR** (to see the entire histogram).

Curve-Fitting (Linear Regression; 2-Variable Statistics)

Example (Using a problem from the Harvard Consortium Calculus Project, 1992):

> Search-and-rescue teams for lost hikers in the West separate and walk parallel to one another. Experience has shown that the percentage P of hikers found is related to the separation distance d of the search team members. Here are some data points:

d (separation distance)	P (% found)
20	90
40	80
60	70
80	60
100	50

This relationship looks linear. Enter the **STAT EDIT** menu and name the xlist DIST and the ylist PERC, pressing **ENTER** (or down cursor ▾) after each name is typed. Then enter the data as respective x's and y's. Next press **2nd**, select **CALC (M1) ENTER ENTER** (or down cursor ▾ twice) and select **LINR (F2)**. The result tells you that the best-fitting *line* for this data is of the form y = a + bx with correlation r. The calculator reports:

LinR
a = 100
b = ⁻.5
corr = ⁻1
n = 5

Here we have

P = 100 − .5d with perfect negative correlation r = ⁻1

If you wish to see the regression equation, press **2nd QUIT** to go to the Home screen, then **2nd RCL**, and finally **STAT VARS (F5) MORE MORE RegEq (F2)**, followed by **ENTER**. The calculator reports

100 + ⁻.5x

To graph the regression line, go to the **GRAPH** menu, deselect any functions in the y(x) = list, and set a suitable **RANGE**, say 0, 200, 10, 0, 100, 1. Then return to **STAT**, select **DRAW (F3)**, and then **DRREG (F4)**. You should obtain the curve:

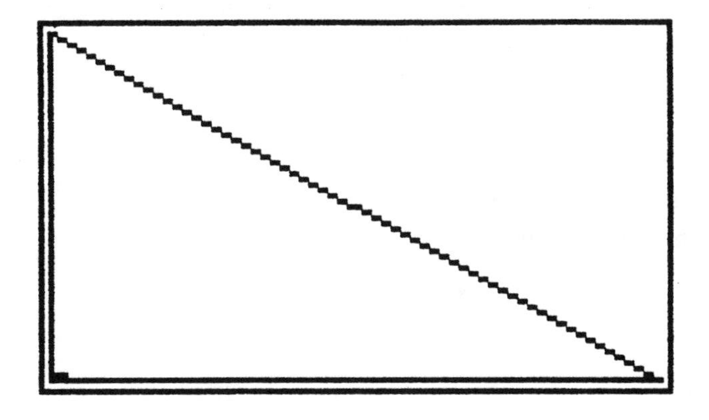

If you wish to interpolate (or extrapolate) to find specific regression-value predictions on this curve, select the **FCST** option, representing "Forecast." It is found in the **2nd M4** menu or by pressing **STAT F4** if you already cleared the menus. Replace either x or y (within the range specified), place the cursor on the desired correspondent, and select **SOLVE (F5)**. For example, a distance of 43 feet corresponds with a prediction of 78.5% recovery rate; a 25% recovery rate corresponds with a separation distance of 150 feet.

Curve-Fitting (Polynomial Regression; 2-Variable Statistics)

Example (From a problem in *Problems for Student Investigation, ACM-GLCA Calculus Reform Project*, Mic Jackson (ed), 1991):

There is a nonlinear relationship between the counter reading (x) of a tape deck and the time (t) that the tape has been playing. Suppose the following data are collected:

counter reading (x)	**time (t)**
0000	0
0374	5
0692	10
0973	15
1227	20
1462	25
1680	30
1886	35
2080	40
2265	45

Find the quadratic polynomial regression fit for this data. That is, find the coefficients a, b, c of

$$t(x) = ax^2 + bx + c$$

Name the variables COUNT and TIME. Use the selection **P2REG** to determine the coefficients, after pressing the sequence of keys

2nd CALC (M1) ENTER ENTER MORE

The calculator reports

{3.44225330619E⁻6, .012069544886, .001264984234}

which means our quadratic regression polynomial is

$$t(x) = .00000344x^2 + .0120695x + .001265$$

You can check this by going to the Home screen, pressing **2nd RCL**, then **STAT VARS (F5) MORE MORE** and selecting **RegEq (F2)**, followed by **ENTER**.

Now, to visualize this fit, go to the **GRAPH** menu and set up a suitable range, such as 0, 2300, 25, 0, 50, 5. Then return to the **STAT** menu, select **DRAW** and then **SCAT** (for scatter plot, **F2**). This displays the 10 data points (x, t) discretely. (The point at the origin is not shown in the figure below.)

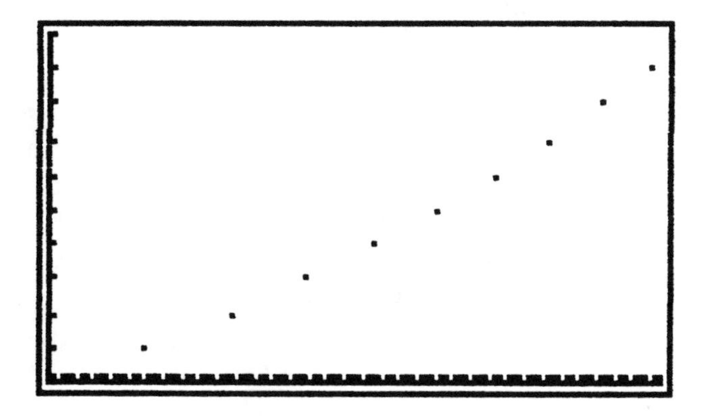

Finally, select **DRREG** (draw regression equation), and the continuous-appearing polynomial function of second degree that best fits the data is drawn. Press **2nd QUIT** to return to the Home screen.

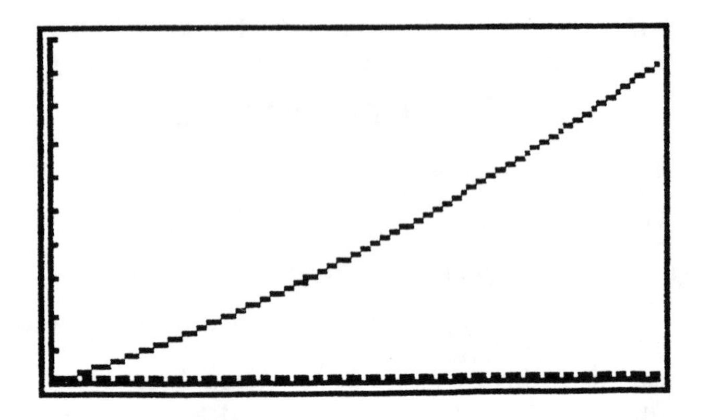

Practice Exercises

1. Find the best-fitting exponential function

$$P(x) = ab^x$$

for the following data sets. Sketch a scatter plot and regression function. You might use the generic names of xStat and yStat under the **STAT EDIT** menu, but before doing so, clear the statistics registers by pressing **CLRxy (F5)** while in the **STAT EDIT** mode. To do the calculation, press **2nd M1 (CALC) ENTER ENTER** and select **EXPR (F4)**.

x	P(x)
$^{-}2$	1.56
$^{-}1$	1.25
0	1
1	0.8
2	0.64
3	0.51
4	0.41

You should obtain:

ExpR
a = .999355183618
b = .799997265815
corr = $^{-}$.999994300996
n = 7

For the scatter plot and regression curve, don't forget to set the GRAPH RANGE first, and deselect any functions in the y(x) = list.

The best-fitting exponential function will be

$$P(x) = (.8)^x$$

2. Suppose you know that a *linear* equation generated the data in the following table:

x	5.2	5.3	5.4	5.5	5.6
y	27.8	29.2	30.6	32.0	33.4

Find the equation both analytically and by linear regression.

Analytical Solution. The equation has the form $y = a + bx$. From the table, each 0.1 increase in x is accompanied by a corresponding 1.4 increase in y. Hence the slope $b = \frac{1.4}{.1}$, or 14. If we extrapolate 52 tenths backwards to $x = 0$, we determine the y-intercept a, by subtracting 52(1.4) from 27.8. Thus $a = -45$. Our line has equation

$$y = -45 + 14x$$

Calculator Solution (By Linear Regression). Press **STAT EDIT (F2)** ▾ ▾ and **F5** (to CLRxy). Then enter the above data. Next, press **2nd M1 (CALC)** ▾ ▾ and select **F2** (LinR). We obtain

LinR
a = $^{-}$45
b = 14
corr = 1
n = 5

3. Now use the data of Exercise 2 to predict the y-coordinate that would accompany an x-value of 10.2 (assuming the linear model holds for a range including this value of x).

   Calculator Solution. Press **2nd M4 (FCST)**, and define x = 10.2 **ENTER**. With the cursor on y = , select **F5 (SOLVE)**, to obtain 97.8 for the result.

4. The population of Mexico from 1980 to 1986 was estimated to be as follows:

Year	**Population (in millions)**
1980	67.38
1981	69.13
1982	70.93
1983	72.77
1984	74.66
1985	76.60
1986	78.59

   Using x to represent the number of years after 1980, and y as the population (in millions), find a regression model that best fits this data and report its parameters.

   Calculator Solution. The most appropriate model is exponential. Here, we obtain:

   ExpR
   a = 67.3806299612
   b = 1.025980189
   corr = .999999946021
   n = 7

   So, the equation is

   $$y = (67.38)(1.026)^x$$

5. Consider the following table, in which the values of three functions f, g, and h are given for certain values of the independent variable x. One of these functions is exponential, another is of the form $y = ax^2$, and one is of the form $y = bx^3$. Which is which?

x	y = f(x)	x	y = g(x)	x	y = h(x)
2.0	4.40	1.0	3.00	0.0	2.04
2.2	5.32	1.2	5.18	1.0	3.06
2.4	6.34	1.4	8.23	2.0	4.59
2.6	7.44	1.6	12.19	3.0	6.89
2.8	8.62	1.8	17.50	4.0	10.33
3.0	9.90	2.0	24.00	5.0	15.49

Answer. $f(x) = (1.1)x^2$ with corr = .999998418008

$g(x) = 3x^3$ with corr = .999991507197

$h(x) = (2.04)(1.5)^x$ with corr = .999999922644

6. Because of improved seeds and modern agricultural techniques, the grain production of a certain geographic region has been increasing. Over a 20-year period, grain production (in millions of tons) was:

Year	1970	1975	1980	1985	1990
Production	5.35	5.90	6.49	7.05	7.64

During this same 20-year period, the population (in millions of people) was also increasing:

Year	1970	1975	1980	1985	1990
Population	53.2	56.9	60.9	65.2	69.7

Fit a linear or exponential function to each set of data. Select whichever fits best. Let t = time in years after 1970.

Calculator Solution.

i. For *production:*

linear regression yields

$$Pr = (5.34) + (.1146)t \quad \text{with corr} = .999934$$

exponential regression yields

$$Pr = (5.38497)(1.01797)^t \quad \text{with corr} = .99895$$

So, a *linear* fit is best here.

ii. For *population*:

linear regression yields

$$P = (52.92) + (.826)t \quad \text{with corr} = .999241$$

exponential regression yields

$$P = (53.19475)(1.01362)^t \quad \text{with corr} = .999994$$

So, an *exponential* fit is best here.

7. Suppose this geographic region in Exercise 6 was self-supporting in grain production in 1970. Was it self-supporting in grain production between 1970 and 1990? What about beyond 1990, assuming these trends continue?

Calculator Solution. Compare ratios of

$$\frac{\text{Production}}{\text{Population}}$$

for each year in the table. (Note that the x- or y-values could be entered as calculations or function values, and the calculator will enter them as numerical approximations to the desired values.) We obtain .1, .104, .106, .108, .110 for each of the five years, so it appears that the region was self-supporting from 1970 to 1990. But if we sketch a graph of

$$y1 = \frac{(5.34) + (.1146)x}{(53.19475)(1.01362)^x}$$

over the x-interval [0, 200] and the y-interval [0, .3], we can see visually that the exponential function in the denominator causes the ratio to decrease eventually to 0.

8. *A Sampling Problem.*
(From the TI-81 manual, adapted here for the TI-85.)

Suppose sampling a number of cities of varying populations with respect to the number of buildings over 12 stories high yields the data:

Population	Number of Buildings
150,000	4
250,000	9
500,000	31
500,000	20
750,000	55
800,000	42
950,000	73

Based on this sample, how many buildings of more than 12 stories would you expect to find in a city of 300,000 people?

Hint: To do this, you will need to test four regression models—linear, logarithmic, exponential, and power.

Calculator Solution.

(a) Enter the data in the statistics register.

(b) Execute (relative to this data) each of the regression models. Record your correlation coefficient r each time. The value of r closest to +1 or ⁻1 indicates the most appropriate model.

(c) Execute the regression model which is best for this data. Note the coefficients.

(d) Return to the Home screen (**2nd QUIT**) and store 300,000 in x. (Use lower-case x by pressing **2nd alpha x-VAR**). Then copy the regression equation to the Home screen by pressing **2nd RCL**, and then pressing **STAT F5 (VARS) MORE MORE** and selecting **F2 (RegEq) ENTER**.

(e) Finally press **ENTER** again (to evaluate the regression equation at x = 300,000), and you should have an answer to the problem.

Calculator Solution. For Part (b),

LINR yields corr = .9573
LNR yields corr = .8982
EXPR yields corr = .9593
PWRR yields corr = .9866

Therefore, power regression best fits this data.

For Part (c), the power regression coefficients are

$$a = 4.87421148034\,\text{E}^{-}8$$
$$b = 1.52941335532$$

So, the regression equation is

$$y = (.0000000487)x^{1.5294}$$

To see this visually, go to the **GRAPH** menu and set the **RANGE** at $[100000, 1000000]_x$ with scale 10,000 and $[0, 100]_y$ with scale 1. Then return to **STAT**, select **F3** for **DRAW**, and select **F4** (for **DRREG**). The regression equation is now drawn over the selected range.

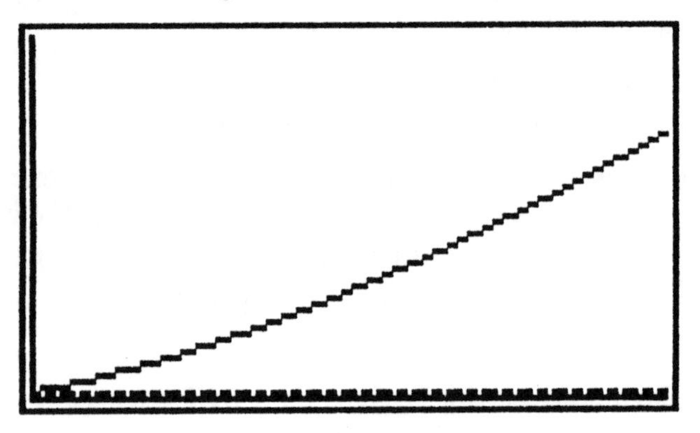

Press **2nd QUIT** to clear the screen and return to the Home screen.

For Part (d), store 300,000 to variable x (lower-case x). Follow the directions in Parts (d) and (e), and we obtain a value of 11.606. This means that (by interpolation) we might expect cities of size 300,000 to have twelve buildings taller than 12 stories.

Note: The results for Parts (d) and (e) could have been obtained from the **STAT FCST** (press **F4**) menu while the power regression model is in the statistics register. Simply enter 300,000 in x, move the cursor ▾ to y (or **ENTER**) and select **SOLVE (F5)**. The calculator indicates 11.606112179107.

Exploration 8 PROGRAMMING THE TI-85

When you press **PRGM** you see two menus: **NAMES** and **EDIT**. To *write* (or change) a program, select **EDIT**, and then either select an existing program named in the submenus or type in a name for a new one and begin creating the program. To *run* an existing program, select **NAMES**, and then select the name of the program of your choice from the submenus.

- To *exit* from a **PRGM** menu, type **2nd QUIT** or **EXIT** (once or twice).

- To *execute* a program:

  1. Exit from the program editor to the Home screen.

  2. Either type in the program's name on the Home screen, or copy the program's name from the **2nd VARS MORE PRGM (F5)** menu, or press **PRGM** and select **NAMES** and then copy the desired name.

  3. Press **ENTER**.

- To *break* during program execution, press and hold **ON** for a moment.

- To *delete* a program:

  Press **2nd MEM**, select **DELET**, then go to the **2nd VARS** menu, press **MORE**, select **PRGM** (**F5**), select the name of the program you wish to delete, place the cursor on its name, and press **ENTER** to delete it.

- **I/O** and **CTL** menus:

  When in **PRGM EDIT**, if you press the **PRGM** button, you see up to 7 submenus. Most programming maneuvers are found in the **I/O** and **CTL** submenus. Note that quotes (") are in the **PRGM I/O MORE** menu, item **F5**.

We begin with a very short (subroutine) program. Press **PRGM EDIT** and type in RANGE for the name of the program. Next, type in the following seven lines, exactly as shown here. Press **ENTER** after each program line, and a colon will appear to define the beginning of the next line.

> *Note:* Input is in the I/O menu, xMin, xMax, and other range settings are in the **GRAPH RANGE** menu. To obtain Input again, press **EXIT**, select **I/O**, and then select **Input** (**F1**). To obtain the **Return** command, press **EXIT 2nd M4** (**CTL**) **MORE MORE** (**F4**).

```
: Input ("xMin  = ", xMin)
: Input ("xMax  = ", xMax)
: Input ("xScl  = ", xScl)
: Input ("yMin  = ", yMin)
: Input ("yMax  = ", yMax)
: Input ("yScl  = ", yScl)
: Return
```

This program, RANGE, will be called from inside another program to demonstrate calling one program from another. It requests user input to set up a suitable viewing window.

Now, exit from the **PRGM** menu (Press **2nd QUIT** or **EXIT** (twice)), and define the function

$$y1 = x^3 - 5x$$

in the **GRAPH y(x)** = menu (or, as a better alternative, input the function *during* program execution). We will now write a program (adapted from Beckman and Sundstrom, *Graphing Calculator Laboratory Manual for Calculus*, Addison-Wesley, 1990). This program requests input about your function, suitable range values, sketches the graph of the function you defined inside your viewing window, and with each press of **ENTER**, it sketches a short tangent line segment to the graph of your function and plots a corresponding point on the graph of the derived function. To begin, return to the **PRGM** menu, select **EDIT** (**F2**), and type in TANGENT in response to Name = . Then copy the following program stepwise. *Don't forget to press* **ENTER** *after each program line. To edit a line, simply scroll* (▲ ▼) *to it and make the necessary corrections.*

PROGRAM: TANGENT

Command	*Location in TI-85*
: ClLCD	in **I/O** (**F3**) **MORE**, option **F3** (clears the screen)
: Func	in **2nd MODE** select **Func**
: FnOff	in **GRAPH VARS (F1) MORE MORE F2**
: ClDrw	in **GRAPH MORE DRAW MORE F4**
: Outpt (3, 1, "Enter y1")	in **I/O** menu (Press **EXIT MORE** and select **F5**; requests user to define function y1 in the row 3, column 1 position. Press **MORE** (when in the **I/O** menu) and select **F5** to obtain quotes (")). The words "Enter y1" must be typed character by character
: Disp " "	in **I/O** menu (Press MORE **F3**; causes a blank line). Press **MORE F5** to get quotes
: InpSt ST1	in **I/O F1** (input a string called ST1)
: St▶Eq (ST1, y1)	**2nd STRNG F5** (converts string ST1 to equation y1)
: RANGE	Type this in caps; it calls the other program called RANGE at this point

: (xMax − xMin)/30 → W	xMax, xMin in **GRAPH RANGE. STO▸** produces →
: 5 → Z	
: DrawF y1	in **GRAPH MORE DRAW F5** (type y1 as 2nd alpha y1)
: Pause	Press **EXIT** (to **CTL** menu) press **2nd M4** (**CTL**) **MORE MORE F3**
: xMin + Z * W → P	*Note:* Use * or space between characters Z, W to indicate multiplication (*not* juxtaposition)
: Lbl LOOP	Press **EXIT** (to **CTL** menu) press **MORE MORE F4** (type in LOOP in all caps)
: P → x	Use lower-case x here
: nDer(y1, x) → M	nDer is in **2nd CALC F2** (then type in y1, x)
: PtOn(P, M)	in **GRAPH MORE DRAW MORE F1**
: Line (P − Z * W, y1 − Z * M * W, P + Z * W, y1 + Z * M * W)	in **GRAPH MORE DRAW** menu (**F2**), or **MORE MORE F2** (from preceding step)
: Pause	in **EXIT** (to **CTL** menu) **MORE F3**
: P + W → P	
: If P + Z * W ≤ xMax	**If** is in press **MORE**, select **F1** (CTL submenu), ≤ is in **2nd TEST** menu, item **F4**; xMax is in **GRAPH RANGE F2**
: Goto LOOP	**EXIT** (to control menu) **MORE F5**, and type in **LOOP**
: Pause	In **CTL** menu, press **MORE** and select **F3**
: Disp "END"	In **2nd M3** (I/O menu), press **F3**; to get quotes, press **MORE** and **F5**
: Stop	In **EXIT**, select **CTL** menu (**F4**) **MORE MORE F5**

Now, in **2nd QUIT PRGM NAMES**, select **TANGE ENTER**, and respond x ^ 3 − 5x to the "Enter y1" question, then $^{-}3$, 3, 0,

$^{-}$10, 10, 0 to the RANGE entries requested. Then press **ENTER** and the curve is drawn. Press **ENTER** again and a tangent segment is drawn and a corresponding point on the derivative graph is plotted, and so on for each successive press of **ENTER**. (After 22 presses, the program returns to the Home screen, reporting END).

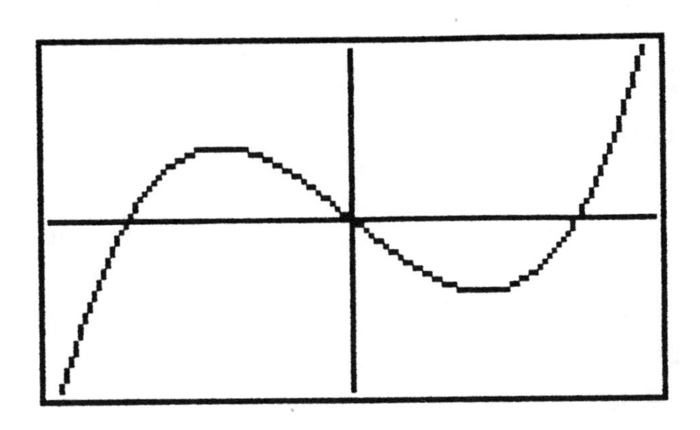

Next, we demonstrate a program (adapted from the TI–85 manual) that illustrates the combination of graphical and numerical perspectives in approximating a root by Newton's Method. In this example, we use the function

y1 = e ^ x − 3x   (input by you during execution)

defined over the range settings (input by you)

$^{-}$3, 3, 0, $^{-}$5, 7, 0

and using a numerical derivative tolerance of δ (input by you).

PROGRAM: NEWTON

: ClLCD	In **I/O MORE** menu. This clears the Home screen
: Func	
: FnOff	
: Outpt (3, 1, "Enter y1")	
: Disp " "	

```
: InpSt ST2
: St▸Eq (ST2, y1)
: RANGE
: Input ("δ = ", δ)                    δ is in 2nd TOLER menu
: Lbl MORE
: Trace                                Trace is in GRAPH menu
: 1 → I
: Lbl TEST
: x − y1/nDer(y1, x) → ROOT
: If abs(x − ROOT) ≤ abs(x/1E10)       Use EE key for 1E10
: Goto DONE
: ROOT → x : I + 1 → I: Goto TEST
: Lbl DONE
: Disp "ROOT = ", ROOT
: Disp "ITER = ", I
: Pause
: ClLCD
: Output (3, 1, "ANOTHER ROOT?")       ? is in 2nd CHAR MISC
                                       (F1) menu
: Disp "  "
: Input R
: If R = = 1                           = = is in 2nd TEST F1
                                       menu
: Then                                 Then is in CTL menu above
                                       F2 key
: Goto MORE
: Stop
```

Now press **2nd QUIT**, go to the Home screen, press **PRGM**, select **NAMES (F1)**, and select **NEWT ENTER**. You will be asked to input your function, then to input all six range entries, in order, and a value for δ (the value of the delta step-size used in calculating numerical derivatives). Respond as follows:

Enter y1
? **e ^ x − 3x**
xMin = **⁻3**
xMax = **3**
xScl = **0**
yMin = **⁻5**
yMax = **7**
yScl = **0**
δ = **.00001**

After you have entered this data and upon pressing **ENTER**, you see a graph of the function

$$y1 = e^x - 3x$$

and a report as to the position of the winking cursor.

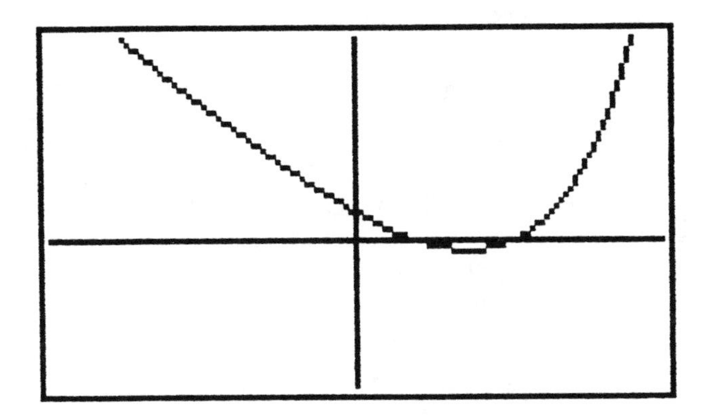

Since the graph has been drawn using a **TRACE** command, move the cursor along the graph to a position near the first root (the one closer to 0).

Using this position as a first guess for Newton's Method, press **ENTER**, and you obtain

ROOT =

.619061286736

ITER =

5 (depends on your cursor guess)

Press **ENTER** to reactivate the program, which asks "ANOTHER ROOT?" If you respond anything but "1", the program ends; if you respond "1" **ENTER**, the curve and trace cursor reappear. This time position the cursor near the other root, press **ENTER** and obtain

ROOT =

1.51213455166

ITER =

4

Press **ENTER** and respond any key but 1 to end the program. Press **CLEAR** to clear the Home screen.

We conclude with a very simple program, which we shall call "VSCREEN". This program enables the user to input a scale factor (for both axes simultaneously) and then sets up the viewing window so that the number of cursor moves in either direction are that scale multiple of .1. It uses the fact that the viewing window is 126 by 62 pixels and divides it into exactly 126 tenths horizontally and exactly 62 tenths vertically.

```
PROGRAM: VSCREEN

: Disp "SCALE?"
: Input S
: ⁻6.3 * S → xMin
: 6.3 * S → xMax
: S → xScl
: ⁻3.1 * S → yMin
: 3.1 * S → yMax
: S → yScl
```

Then press **2nd QUIT** to return to the Home screen.

This program is especially handy for demonstrating removable (missing point) discontinuities of certain functions. For example, go to the **GRAPH** menu and define the function

$$y1 = (x^2 - 1)/(x - 1)$$

Next, instead of setting up a specific range, press **2nd QUIT**, go back to the **PRGM** menu, press **NAMES**, and select **VSCRE**, **ENTER**. Now you are asked to input the scale factor

SCALE?

Respond 1, **ENTER**, and the value of 1 appears on the Home screen. The viewing window (range) is now set at $[^-6.3, 6.3]_x$ and $[^-3.1, 3.1]_y$ with scale factors 1. Press **GRAPH** and then **F5** (Graph), and the graph of the line y = x + 1 with a hole at (1, 2) is shown clearly. Note that if you move the cursor around the screen, the x- and y-coordinates change by exactly .1 unit for each move. If you select **TRACE** and move the cursor along the graph to the hole, the calculator reports a value of 1 for x and no value for y, since the function is not defined there.

There are many different program commands and options in the TI-85. We have only "scratched the surface" with these examples.

Practice Exercises

1. Modify the **TANGENT** program so that it only plots points (function values, not tangent lines) for the derivative function along with the graph of the desired function. Also, modify it so that the points plotted are more closely spaced.

   Calculator Solution. For closer spacing, change the line "5 → Z" to "2 → Z." To remove the drawing of tangent lines and plot only points, clear the command that begins with

   Line (

   Additionally, if you wish to see more points on the derivative function, change the line

   If P + Z * W ≤ xMax

   to

   If P ≤ xMax

2. Try out your modified TANGENT program in the following situations:

   (a) First define

   $$y1 = 2 \wedge x$$

   with range settings $^{-}2$, 2, 0, $^{-}1$, 4, 0. Then, press **ENTER** slowly and successively about 45 times. You should see on the screen:

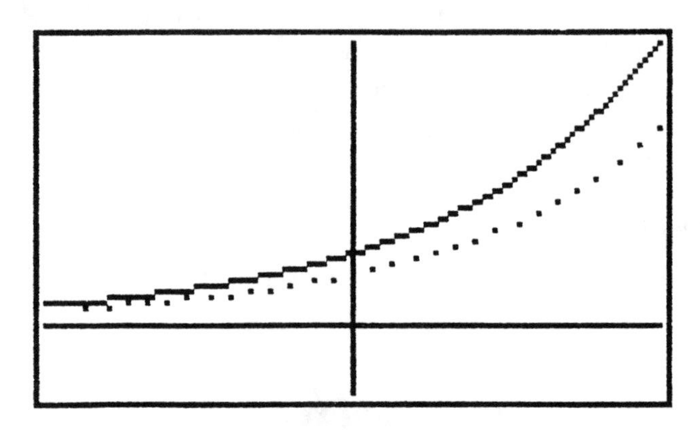

   (b) Next press **ENTER** again to rerun the program, and define

   $$y1 = 3 \wedge x$$

   with range settings $^{-}0.5$, 2, 0, $^{-}1$, 9, 0. Be patient while pressing **ENTER** successively, about 45–50 times. Some of the dots may merge with the "continuous" curve.

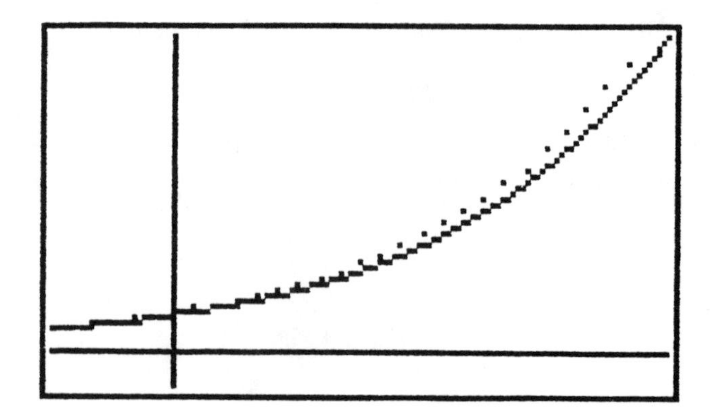

(c) Why are the dots below the curve $y = 2^x$ and above the curve $y = 3^x$? Under what condition will the dots be exactly *on* the curve?

Answer. The derivative of

$$y = 2^x$$

is

$$y' = 2^x \ln 2$$

where $\ln 2$ is a number smaller than 1. So the derivative (dotted) curve is pulled downward slightly from the original function curve. On the other hand, the derivative of

$$y = 3^x$$

is

$$y' = 3^x \ln 3$$

where $\ln 3$ is a number slightly larger than 1. Hence, the derivative (dotted) curve in this case is slightly above the original function curve. The function that exactly matches its own derivative is an exponential function having a base between 2 and 3. This function is

$$y = e^x$$

3. Use the Newton program to approximate all roots of the function

$$y = 2x^3 - x^2 - 10x + 5$$

Set up the range with x- and y-scales of 0 and $[^-3, 3]_x$ and $[^-10, 20]_y$. Use δ tolerance at .001. Don't forget to respond "1" to the question "ANOTHER ROOT?" if you wish to search for additional roots.

Answer.

$$^-2.2360679775, \quad .5, \quad 2.2360679775$$

(Compare these answers with those of Exercise 3 in Exploration 2.)

4. Use the Newton program to approximate the two positive roots of the function

$$f(x) = \frac{x^3 - 10x^2 + x + 50}{x - 2}$$

Use a viewing window of $[^{-}10, 10]_x$ and $[^{-}100, 100]_y$ with x- and y-scales of 0 and $\delta = .001$. (Before you run the program, you may wish to set the **GRAPH MORE FORMT** to the **DrawDot** option.)

Answer.

$$2.68337520964, \quad 9.31662479036$$

(Compare these answers with those of Exercise 4 of Exploration 3.)

5. Use the Newton program to find a nonzero solution to the equation

$$x + 2x \ln x = 0$$

Answer.

$$x \approx .606530659713$$

6. What does the answer in Exercise 5 have to do with the solution to the nerve-impulse-transmission problem in Exploration 3?

Answer. In the nerve-impulse-transmission problem, the function to be maximized was

$$y = -x^2 \ln x$$

The derivative of this function is

$$-(x + 2x \ln x)$$

If we set the derivative equal to zero and solve for x, we have the situation we have just shown in Exercise 5.

7. Consider the function

$$y = \frac{(x-2)(x+3)(x^2-5)}{(x-2)(x+3)}$$

Go to the **GRAPH** menu. Make sure the **GRAPH MORE FORMT** is reset to the **DrawLine** option. Define the given function above in the y(x) = menu as y1. Then **2nd QUIT** from the GRAPH menu to the Home screen, go to the **PRGM** menu, select **NAMES**, and select the **VSCRE** program. Set the scale at 2.5. Then return to the **GRAPH** menu, check the **RANGE**, and **GRAPH** (**F5**) the function. What do you see, and why is the range as it is?

Calculator Solution. What you see is the parabola

$$y = x^2 - 5$$

with two holes in it at $x = 2$ $(y = -1)$ and $x = -3$ $(y = 4)$. (Use **TRACE** to verify this.) The range settings are due to a scale factor of 2.5 (input by you) that multiplies the x-scale [⁻6.3, 6.3] and y-scale [⁻3.1, 3.1] by this factor.

Exploration 9 MISCELLANEOUS OPERATIONS

Polar Equation Graphing

Graph the limaçon

$$r = 2 + 3 \cos \theta$$

First, set the mode to Polar Graphing by pressing **2nd MODE**; select **Pol** in the function section. (Make sure the **RADIAN** setting is engaged.) Then **EXIT GRAPH MORE**, and select **FORMT**; then move the cursor to **PolarGC**. Also, use the **DrawLine** setting (for better graphs). Now press **F1** (the r(θ) = menu) and define your function by responding (to r1 =) **2 + 3 cos θ** (θ is in the submenu at **F1**) **ENTER**. Next, set an appropriate range: **2nd M2** (RANGE) and define

$$0, 2\pi, \pi/24, ^{-}6, 6, 0, ^{-}6, 6, 0$$

as the settings. Then press **F5** (GRAPH). [See the figure on the top of page 77. If you wish a square viewscreen, select **ZOOM MORE** and select **ZSQR**.]

Now, to *store* this graph and its database (all range settings, modes, formats, and equations), press **EXIT** (to the GRAPH menus) **MORE** and select the option **STGDB**; then give your

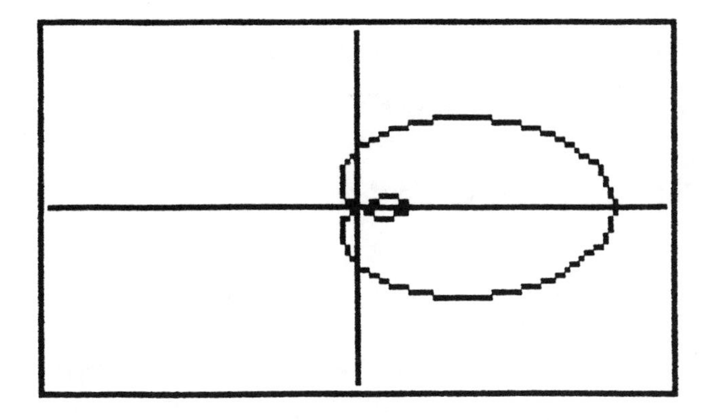

graph database a name, say LIM. When you press **ENTER**, this graph database is stored under **2nd VARS MORE MORE GDB**. (Check to see.) If you wish to recall it later, return to the **GRAPH** menu, **MORE**, select **RCGDB** (for recall graph database), and select the submenu in which LIM appears. From the Home screen or in a program, simply type **RCGDB LIM** character by character and the graphing and format modes will reset for this database, and then when you enter the **GRAPH** menu, you can recall it quickly. Other graphing operations, like TRACE, ZOOM, DRAW, EVAL, and MATH work much the same as they did in Cartesian function graphing mode. Press **2nd QUIT** or press **EXIT** once or twice to exit from the existing menus.

Pairs of Parametrically Defined Equations

This works in an analogous fashion to polar graphing. First the **2nd MODE** must be set to **Param**; then **2nd QUIT** to the Home screen and press **GRAPH**. The equations are entered in *pairs* after selecting **E(t) =** (**F1**). The RANGE has 3 settings each for the parameter t and for the dependent variables x and y. For example, to graph the ellipse

$$x = 3 \cos t$$
$$y = 5 \sin t$$

press **2nd MODE** and select **Param ENTER GRAPH** and menu option **E(t)=** (above **F1**). Now respond

xt1 = **3 cos t** **ENTER** (t is in menu option **F1**)
xt2 = **5 sin t** **ENTER**

then **2nd M2** (for RANGE), and set

$$0, 2\pi, \pi/24, {}^{-}3, 3, 0, {}^{-}5, 5, 0$$

for the range settings. Press **F5 (GRAPH)** and you have your ellipse. The other graphing operations behave much like polar and function graphing. For example, while inside the **GRAPH** menu, press **MORE** and select the **MATH** submenu. Suppose you wish to find the slope of the tangent line, dy/dx, at some point on this ellipse. Select **dy/dx (F2)**; you obtain a report of the cursor location at $t = 0$, $x = 3$, $y = 0$. Now move the cursor (using the ▸ key) to

$$t = .6544984695,\ x = 2.3800600209,\ y = 3.043807145$$

and press **ENTER**. The calculator reports

$$dy/dx = {}^{-}2.172042288$$

To exit these menus, press **2nd QUIT**. You may also wish to reset **2nd MODE** to **Func**. Then press **2nd QUIT** to return to the Home screen.

Vector Operations

To enter a vector from the Home screen, use *square* brackets (**2nd[** or **2nd]**), with components separated by commas. Vector operations are located in the **2nd VECTR** menu.

Examples:

1. Find a unit vector in the direction of [3,4] and return its components as fractions.

   Calculator Solution. **2nd VECTR**, select **MATH (F3)**, then **UnitV (F2)** and type in **[3, 4]** and **2nd MATH MISC MORE** and select **▸ Frac (F1) ENTER**. Obtain

   [3/5 4/5]

Note: If the resulting components are irrational numbers, then the ▶Frac command is inoperative. Try, for example, unitV [2, 3] ▶Frac. The result involves only decimal approximations, not exact arithmetic.

2. Find the dot product of [1, 2, 3] with [$^{-}1$, 5, 7].

   Calculator Solution. **2nd VECTR MATH (F3)**, select **dot (F4)**, and then type in **[1, 2, 3], [$^{-}$1, 5, 7])** **ENTER,** and receive a result of 30.

3. Find the cross product (vector) of [1, 2, $^{-}1$] × [$^{-}2$, 2, 2].

   Calculator Solution. **2nd VECTR MATH (F3)**, select **cross (F1)**, and type in the vectors, so that you have

   cross ([1, 2, $^{-}$1], [$^{-}$2, 2, 2])   ENTER

   The calculator yields [6 0 6].

4. Convert the vector [1, 2, 3] to *spherical-coordinate* representation.

   Calculator Solution. Type in **[1, 2, 3]** on the Home screen, then **2nd VECTR OPS (F4)**, and select **▶Sph (F5) ENTER**. You obtain (after scrolling)

   [3.74165738677 ∠ 1.10714871779 ∠ .640522312679]

   If you wish to return to the original (Cartesian) vector, press **MORE** and select **▶Rec (F1) ENTER**, and you convert back to [1, 2, 3]. Press **EXIT** (twice), then **CLEAR** to clear the screen.

Complex Numbers

Complex numbers of the form a + bi are entered by using *round* parentheses and components separated by a comma. Complex operations are in the **2nd CPLX** menu.

Example:

Consider the complex number 2 + 3i. Find (a) its conjugate, (b) its real part, (c) its imaginary part, (d) its magnitude, (e) its polar angle, and (f) its polar representation.

Calculator Solution. First, let's store (2, 3) as a single-letter complex variable. Type **(2, 3) → A** and press **ENTER**. Then press **2nd CPLX** and do the following:

(a) Select **conj (F1)**, **ALPHA A ENTER**, to obtain (2, ⁻3).

(b) Select **real (F2)**, **ALPHA A ENTER**, to obtain 2.

(c) Select **imag (F3)**, **ALPHA A ENTER**, to obtain 3.

(d) Select **abs (F4)**, **ALPHA A ENTER**, to obtain

3.60555127546

(e) Select **angle (F5)**, **ALPHA A ENTER**, to obtain

.982793723247 (in radians)

(f) Press **MORE**, press **ALPHA A**, select **▸Pol**—to obtain

(3.60555127546 ∠ .982793723247)

Press **EXIT**, then **CLEAR** the Home screen.

Conversions

The TI-85 has built-in conversions (use the **2nd CONV** menu) for 11 different physical-measurement quantities, with submenus containing varieties of different appropriate units for each quantity.

Examples:

1. Convert 15 square feet to square inches.

Calculator Solution. Press **2nd CONV**, select **AREA (F2)**, type in 15, and select **ft²** **(F1)**. Press **MORE** and select **in²** **(F1)** to obtain a display of

$$15\ \text{ft}^2 \blacktriangleright \text{in}^2$$

Then press **ENTER** to obtain 2160.

2. Convert 60 mi/hr to ft/sec.

   Calculator Solution. Press **2nd CONV MORE MORE** and select **SPEED (F1)**. Then type 60, select **mi/hr (F3)**, select **ft/s (F1)**, **ENTER** to obtain 88.

Press **EXIT** once to return to the conversion menus; press **MORE** to access the **VOL** conversion for the next example.

3. How many liters per second are flowing in a pipe in which the fluid is moving at a rate of 32,185 gallons per hour?

   Calculator Solution. The units must both be of the same "conversion type." This problem (and problems like it) must be done in several stages.

   Convert 32185 gallons to liters, obtaining 121833.478268. Then temporarily store this result in a variable, say A. Next, convert hours to seconds obtaining 3600, and store this in B. Then divide A by B and obtain 33.8426328522.

4. Convert 57.3215° to degrees, minutes, and seconds.

   Calculator Solution. This kind of conversion is handled in the **2nd MATH** menu, **ANGLE** submenu. From the Home screen, type 57.3215 **2nd MATH ANGLE** and select **F4** (**▸DMS**) **ENTER**. The calculator indicates 57° 19′ 17.4″. (Note that the DMS conversion assumes the number input to be in degrees). Press **EXIT** (twice) and **CLEAR** the screen.

Inverses of Functions

Here we consider how to graph a function and its inverse (relation).

In the **GRAPH** menu, define y1 = tan x and set the **RANGE** at ⁻2, 2, 0, ⁻2, 2, 0. Also press **MORE**, select **FORMT**, and select the **DrawDot** option. Now press **GRAPH (F5)** and sketch the graph of the function. Next press **MORE**, select **DRAW (F2)**, then **MORE MORE**, and select **DrInv (F2)**. Return to the Home screen and type in **2nd alpha Y 1 ENTER**. Now you see both the tangent function and its inverse ($\tan^{-1}$) on the same viewing screen.

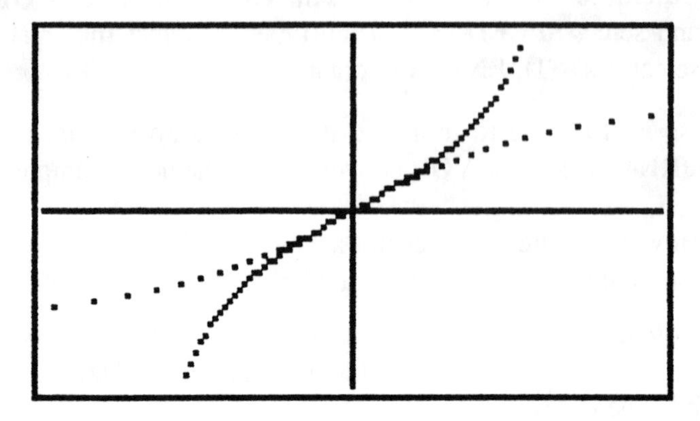

Try this with functions that have inverses (that are *functions*), and other functions that have *inverse relations.*

Examples:

1. Consider

$$f(x) = \frac{x^5 + 7}{10}$$

over $[^-2, 2]_x$ and $[^-2, 2]_y$, where

$$f^{-1}(x) = (10x - 7)^{1/5}$$

is its inverse *function.* Follow the given instructions to graph both f and f^{-1}.

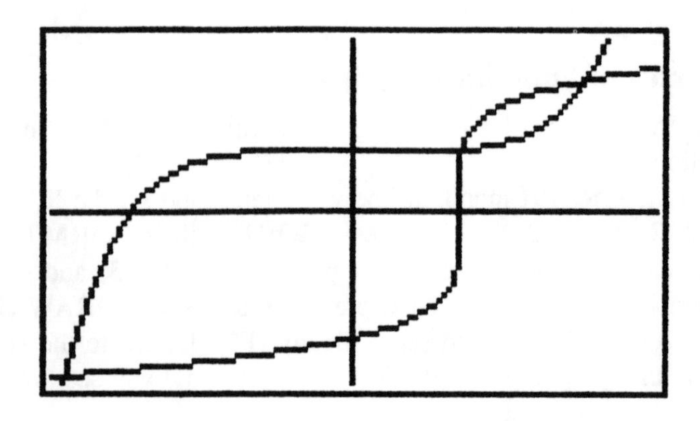

In Examples 2 and 3, the inverses of the functions are relations.

2. Consider

$$g(x) = x^5 - 3x$$

over $[^-3, 3]_x$ and $[^-3, 3]_y$. Sketch the graphs of g and g^{-1}.

Calculator Solution. Hint: First set **GRAPH FORMT** to **DrawLine.**

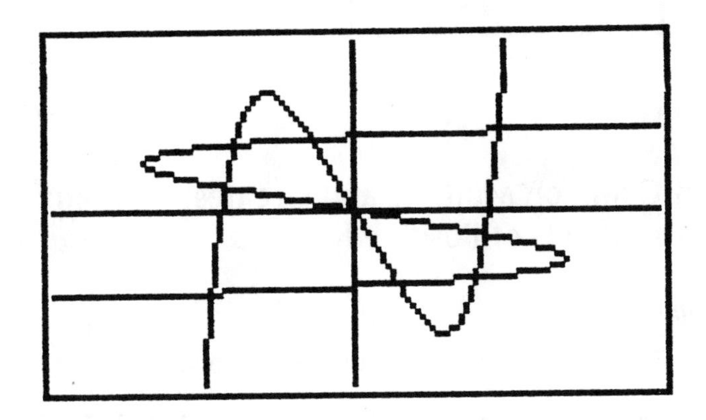

3. Try the sin x function over $[^-2\pi, 2\pi]_x$ and $[^-2\pi, 2\pi]_y$.

Calculator Solution. After defining the function in the y(x) = menu and setting the given range values, use **ZOOM MORE**, select **ZSQR (F2)** to draw the original graph. Then press

MORE DRAW (F2) MORE MORE DrInv (F2)

and type

2nd alpha Y 1 ENTER

to complete the graph, as shown on page 84. Press **2nd QUIT** to return to the Home screen.

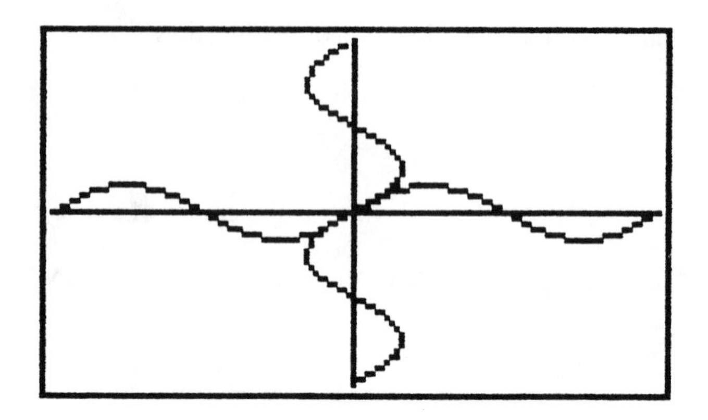

Least Common Multiple and Greatest Common Divisor of Sets of Positive Integers

Examples:

1. Find the least common multiple of 72 and 60.

   Calculator Solution. This is in the **2nd MATH MISC (F4)** menu. Select **lcm (F4)** and type in **72, 60) ENTER**. You obtain 360 as the result.

2. Find the lcm of 72, 60, and 14.

   Calculator Solution. The lcm operation only acts on two arguments at a time, so this is accomplished by going to the **2nd MATH MISC (F4)** menu, selecting **lcm (F4)**, and typing

   lcm (lcm (72, 60), 14) ENTER

   The calculator returns 2520 as the result.

3. Find the greatest common divisor of 72 and 60.

   Calculator Solution. In the **2nd MATH MISC** menu, select **gcd (F5)**, and type **72, 60) ENTER**. The result is 12.

4. Find the greatest common divisor of 72, 60, 42, and 198.

   Calculator Solution. Use the sequence

   gcd(gcd(72, 60), gcd(42, 198))   ENTER

   and obtain 6. Press **EXIT** (twice), then **CLEAR** to clear the screen.

Roots of Polynomials

The TI–85 will find all real and complex roots of polynomials having order up to 30. This is handled in a special menu, **2nd POLY** (above the PRGM key), and uses the **SOLVER** routine. The user responds to the order of the polynomial and then inputs the coefficients *in descending order.* Then, selecting the option **SOLVE** (**F5** key), the roots are calculated.

Examples:

1. Solve the quadratic equation

   $$x^2 + 5.5x - 3 = 0$$

   Calculator Solution. Press **2nd POLY**, respond **2 ENTER** to order =. Then place the coefficients as

   a2 = 1   **ENTER**
   a1 = 5.5   **ENTER**
   a0 = ⁻3   **ENTER**

   and select **SOLVE** (in the **F5** key). The calculator indicates:

   x1 = ⁻6
   x2 = .5

2. Solve the quartic equation

   $$x^4 + 1 = 0$$

   Calculator Solution. Press **2nd POLY**, respond **4 ENTER** to order =. Then respond 1, 0, 0, 0, 1 (each followed by

ENTER) to the respective coefficients a4, a3, a2, a1, and a0. Next select **SOLVE** (**F5**) and obtain the four complex roots:

$$\begin{aligned}
x1 &= (^{-}.707106781186,\ .707106781186)\\
x2 &= (^{-}.707106781186,\ ^{-}.707106781186)\\
x3 &= (.707106781187,\ .707106781187)\\
x4 &= (.707106781187,\ ^{-}.707106781187)
\end{aligned}$$

Note: If you want to solve

$$x^4 - 1 = 0$$

just press **COEFS** (key **F1**) and this returns you to the preceding list of coefficients. Then adjust a0 to be $^{-}1$ and select **SOLVE** (**F5**). As expected, you obtain two real roots ($^{-}1$, 0) and (1, 0) and two complex roots (0, 1) and (0, $^{-}1$). [Actually, the first entries are approximations to zero on the order of 10^{-14}.]

3. Solve the sixth-degree polynomial

$$x^6 - 21x^5 + 175x^4 - 735x^3 + 1624x^2 - 1764x + 720 = 0$$

This was contrived from

$$(x - 1)(x - 2)(x - 3)(x - 4)(x - 5)(x - 6) = 0$$

Calculator Solution. Using the **2nd POLY** menu and responding appropriately to the order and descending-coefficient requests, we obtain (upon pressing **SOLVE**):

$$\begin{aligned}
x1 &= 6.00000000018\\
x2 &= 4.99999999945\\
x3 &= 4.00000000065\\
x4 &= 2.99999999965\\
x5 &= 2.00000000008\\
x6 &= .999999999996
\end{aligned}$$

Press **2nd QUIT** to return to the Home screen.

The Cycloid Family (Parametric Graphing)

We return to parametric graphing to display certain variations in the cycloid family of curves. Press **2nd MODE** and select **Param**. In each case we discuss the physical nature of the curve, provide a general parametric description, and give a specific pair of equations with appropriate range settings to try out. Here, all *scale* settings will be 0, unless otherwise specified.

Cycloids. A cycloid is the path of a point on a circle rolling along a line.

The general equations for the cycloid are

$$x = at - a \sin t$$
$$y = a - a \cos t$$

where a is the distance (radius) from the center of the circle to the moving point.

Sketch the graph of the cycloid:

$$\mathbf{xt1 = 2t - 2 \sin t}$$
$$\mathbf{yt1 = 2 - 2 \cos t}$$

Use $[0, 25]_t$ with t-Step .5, $[0, 50]_x$ and $[^{-}10, 10]_y$ (scales 0).

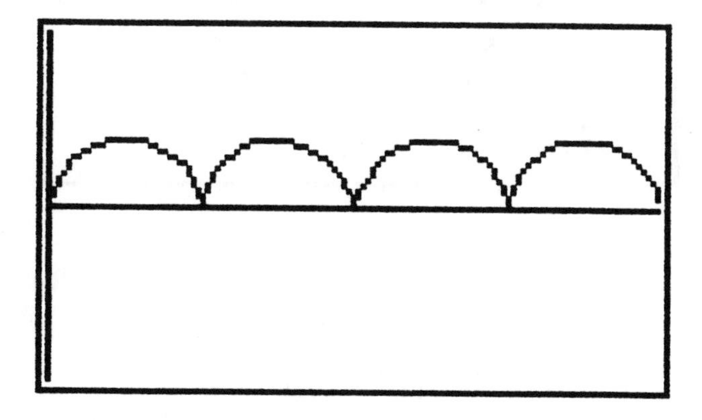

Hypocycloids. If one circle of radius b rolls *inside and tangent to* another circle of radius a (where a > b), and we watch a specific point on the rolling circle, its path is a *hypocycloid.* If a is an integer multiple of b (say a = nb), we obtain a hypocycloid of n cusps. Hypocycloids are also known as *astroids* or *star-shaped curves.*

The general equations for the hypocycloid are:

$$x = (a - b)\cos t + b\cos\left(\left(\frac{a-b}{b}\right)t\right)$$

$$y = (a - b)\sin t - b\sin\left(\left(\frac{a-b}{b}\right)t\right)$$

Sketch the graph of the hypocycloid:

$$xt1 = \mathbf{5\cos t + \cos(5t)}$$
$$yt1 = \mathbf{5\sin t - \sin(5t)}$$

Use $[^-3.14, 3.14]_t$ with t-Step .2, $[^-9, 9]_x$ and $[^-6, 6]_y$. Here a = 6 and b = 1. So we have a hypocycloid of 6 cusps.

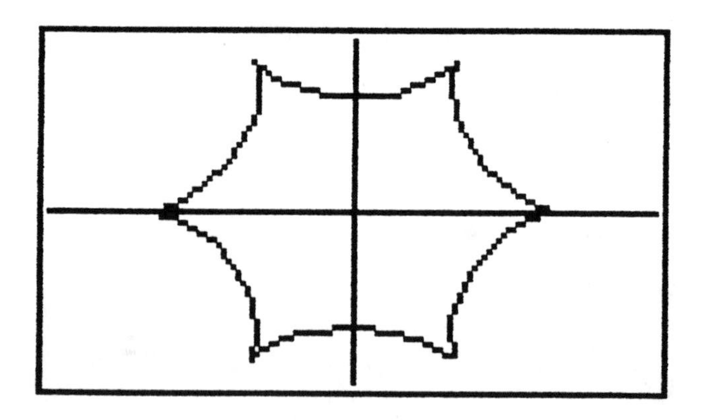

Now graph the circle of radius 6 using the same **RANGE** settings. Leave the hypocycloid in place and define the additional parametric set:

$$xt2 = \mathbf{6\cos t}$$
$$yt2 = \mathbf{6\sin t}$$

Try out different variations, such as

i. a = 6, b = 1.5 (4-cusped hypocycloid)

$x = (4.5)\cos t + (1.5)\cos(3t)$

$y = (4.5)\sin t - (1.5)\sin(3t)$

Use the **E(t)** = menu to define these as parametric functions xt1 and yt1.

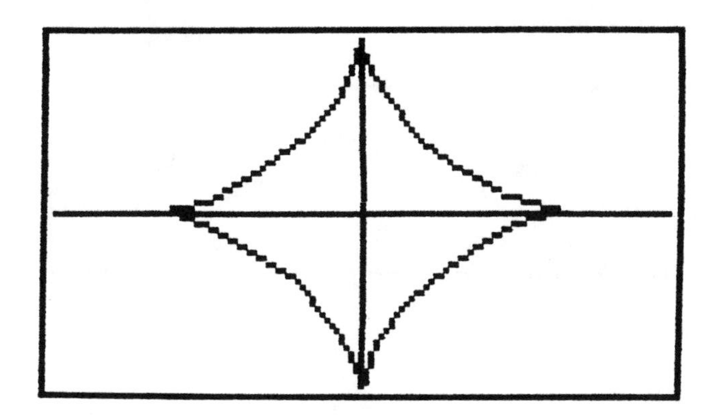

ii. a = 6, B = 2.5 (a is not an integer multiple of b)

$$x = (3.5)\cos t + (2.5)\cos\left(\left(\frac{3.5}{2.5}\right)t\right)$$

$$y = (3.5)\sin t - (2.5)\sin\left(\left(\frac{3.5}{2.5}\right)t\right)$$

Here, let t range from $^{-}6\pi$ to 6π, and leave the other settings as they are.

Epicycloids. If one circle of radius b rolls *outside and tangent to* another circle of radius a (where a > b), and we watch a specific point on the rolling circle, its path is an *epicycloid.* If a is an integer multiple of b (say a = nb), we obtain an epicycloid of n cusps.

The general equations for the epicycloid are:

$$x = (a + b)\cos t - b\cos\left(\left(\frac{a + b}{b}\right)t\right)$$

$$y = (a + b)\sin t - b \sin\left(\left(\frac{a + b}{b}\right)t\right)$$

Sketch the graph of the epicycloid of 6 cusps:

$$\mathbf{xt1 = 7 \cos t - \cos (7t)}$$
$$\mathbf{yt1 = 7 \sin t - \sin (7t)}$$

Use $[^-3.14, 3.14]_t$ with t-Step .2, $[^-15, 15]_x$ and $[^-10, 10]_y$ with scales of 0. Here a = 6 and b = 1.

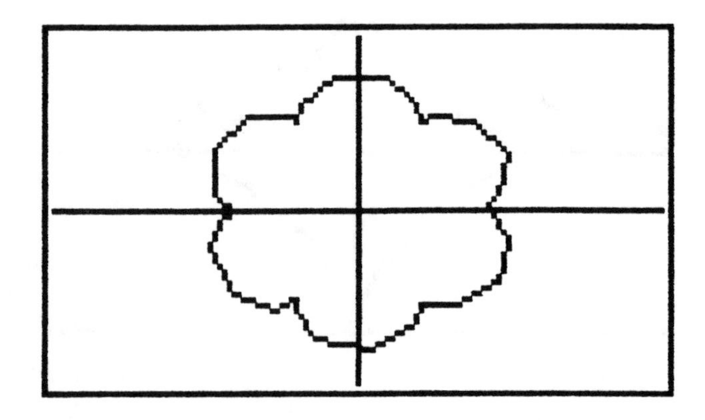

Now graph the circle of radius 6 using the same RANGE settings. The parametric form of the equation is

$$x = 6 \cos t, \qquad y = 6 \sin t$$

Curtate Cycloids. This is a variation on the standard cycloid, where the moving point (whose distance to the center is a) is *closer* to the center of the rolling circle than its radius distance (b). (Think of the motion of a valve stem on an automobile tire as the tire rolls along a level street.)

The general equations of the curtate cycloid are:

$$x = bt - a \sin t$$
$$y = b - a \cos t, \qquad \text{where } a < b$$

Sketch the graph of the curtate cycloid

$$\mathbf{xt1 = t - .75 \sin t}$$
$$\mathbf{yt1 = 1 - .75 \cos t}$$

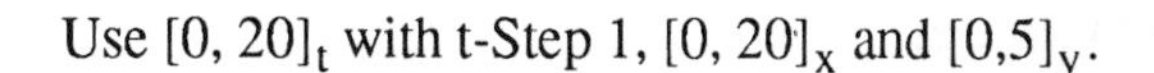

Use $[0, 20]_t$ with t-Step 1, $[0, 20]_x$ and $[0,5]_y$.

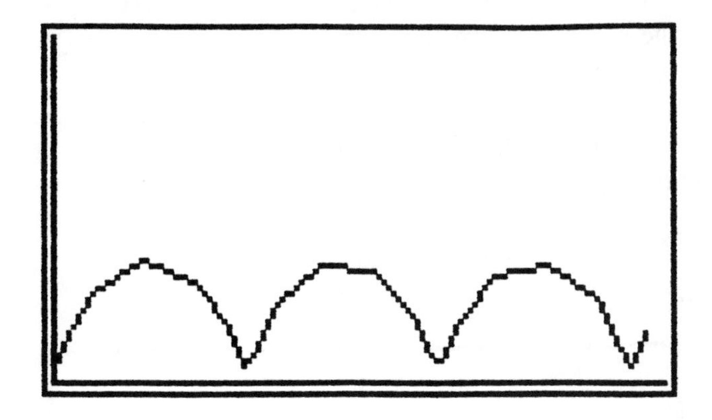

Prolate Cycloids. In this variation on the standard cycloid, the moving point is situated *farther away* from the center of the rolling circle than 1 radius length. (Think of a point on the flange of a railroad wheel, where the rolling circle is tangent to the track.)

The general equations of the prolate cycloid are the same as those for the curtate cycloid, except here a > b.

Sketch the graph of the prolate cycloid

$$\mathbf{xt1 = t - 1.5 \sin t}$$
$$\mathbf{yt1 = 1 - 1.5 \cos t}$$

Use $[0, 25]_t$ with t-Step .5, $[0, 25]_x$ and $[^-3,3]_y$.

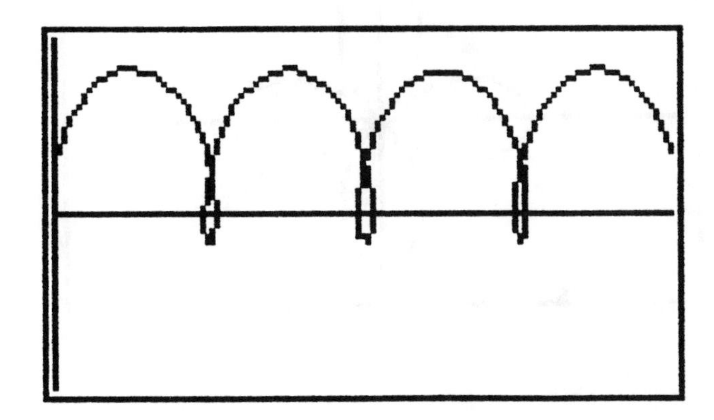

Additional Polar Curves

Set **2nd MODE** to **Pol** and examine some of these curves using the range settings suggested. Also **ZOOM MORE ZSQR** helps improve some pictures. Use x- and y-scale settings of 0 unless otherwise indicated. Some of the following figures are produced with the ZOOM SQUARE option; others are not. The settings appear on page 93.

Roses.

i. $r1 = \mathbf{5}\ \mathbf{sin}(\mathbf{3}\theta)$ (3 *petals*)

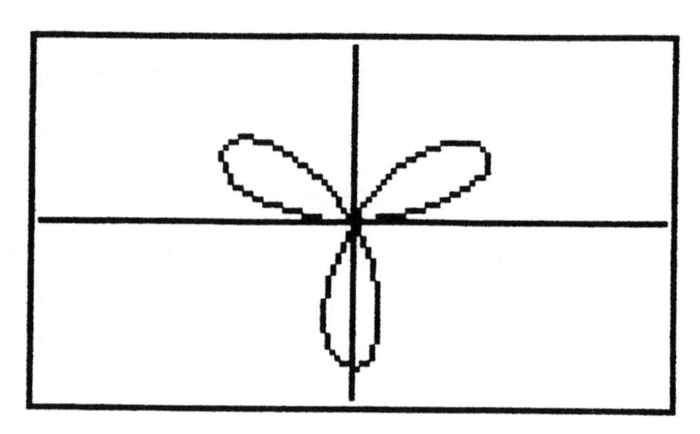

ii. $r2 = \mathbf{5}\ \mathbf{cos}(\mathbf{2}\theta)$ (4 *petals*)

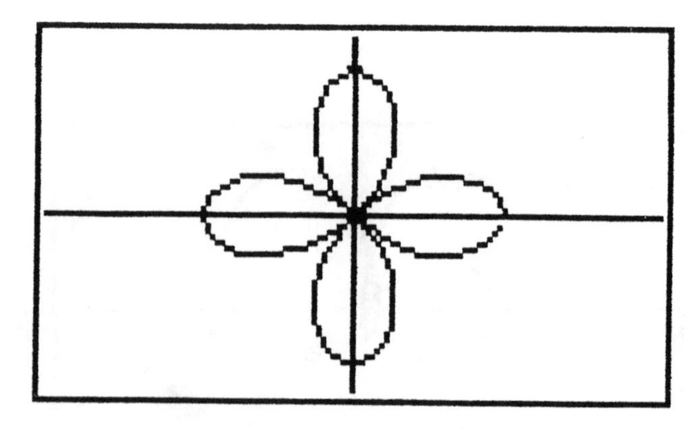

Use $[0, \pi]_\theta$ for r1, $[0, 2\pi]_\theta$ for r2, θ-step .1, $[^-6, 6]_x$ and $[^-6, 6]_y$ for both. (Deselect r1 to graph only r2.)

Cardioids.

$$r1 = 3 + 3 \sin \theta$$

with $[0, 2\pi]_\theta$, θ-step .1, $[^-7, 7]_x$ and $[^-7, 7]_y$. Use **ZOOM MORE ZSQR** (**F2**).

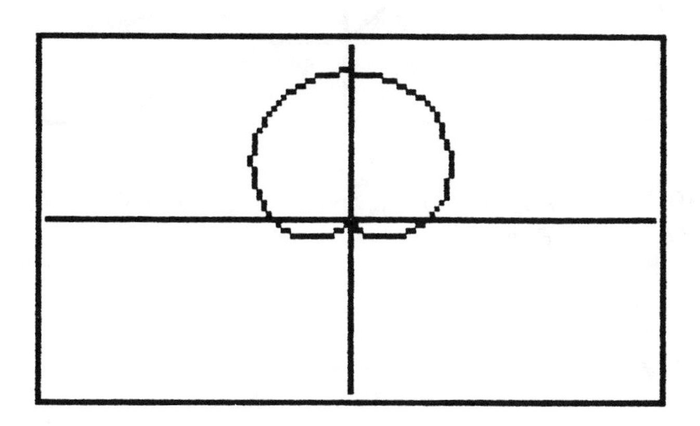

Limaçons.

i. $$r1 = 2 + 4 \cos \theta$$ (*loop*)

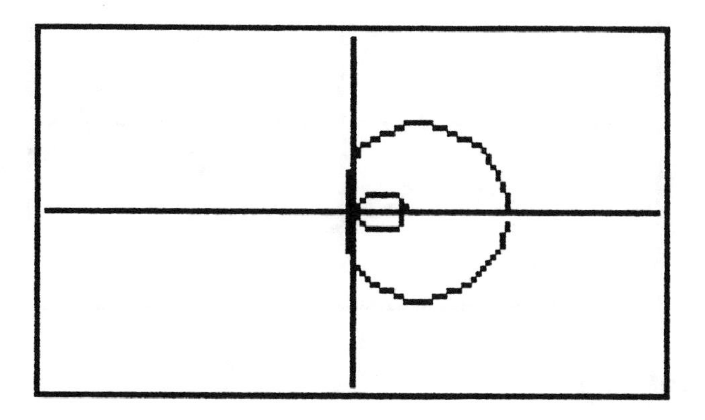

ii. r2 = **4 + 2 cos θ** *(dent)*

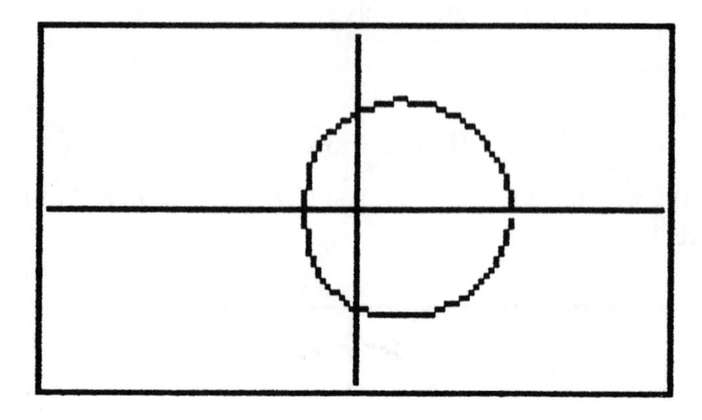

Use $[0, 2\pi]_\theta$, θ-step .1, $[^-7, 7]_x$ and $[^-7, 7]_y$ for both; then use the ZOOM SQUARE setting.

Spirals.

i. r1 = θ *(Archimedes)*

with $[0, 4\pi]_\theta$, θ-step .1, $[^-12, 12]_x$ and $[^-8, 8]_y$.

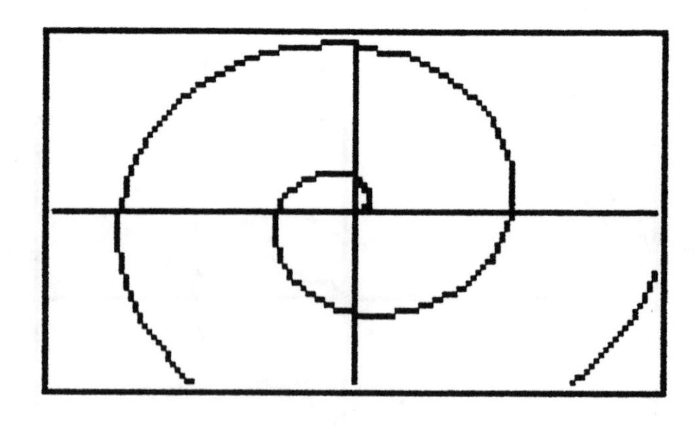

ii. r1 = **e ^ θ** *(logarithmic)*

with $[0, 3]_\theta$, θ-step .1, $[^-9, 9]_x$ and $[0, 9]_y$.

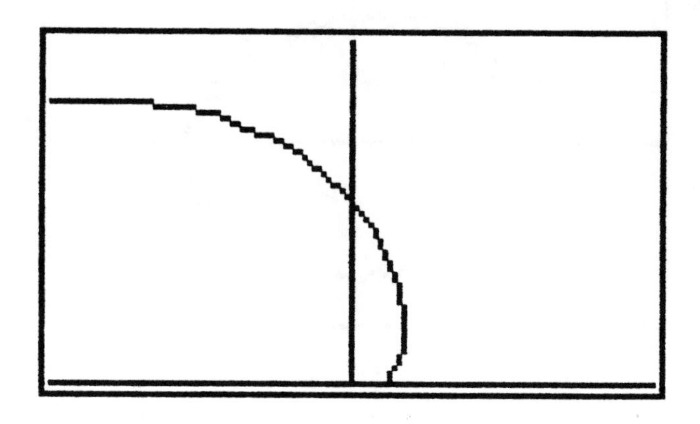

iii. $r1 = 1/\theta$ (*hyperbolic*)

with $[^-2\pi, 2\pi]_\theta$, θ-step .1, $[^-1, 1]_x$ and $[^-1, 1]_y$.

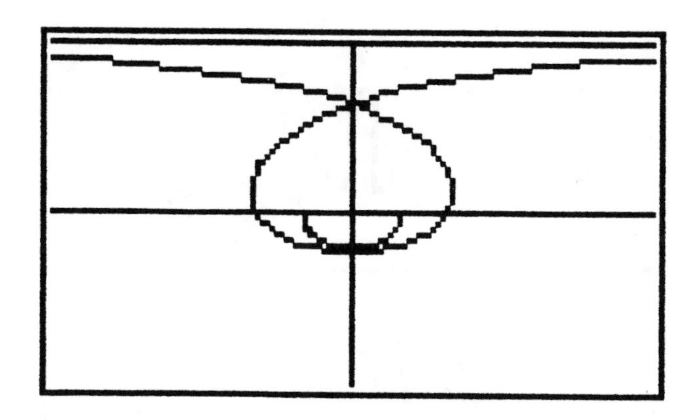

Circular Saw Blades.

$$r1 = \mathbf{2 + 8\theta/\pi - (int(16\theta/\pi))/2}$$

with $[0, 2\pi]_\theta$, θ-step .1, $[^-4, 4]_x$ and $[^-3, 3]_y$—(int is in **2nd MATH NUM** (**F1**), then use **F4**).

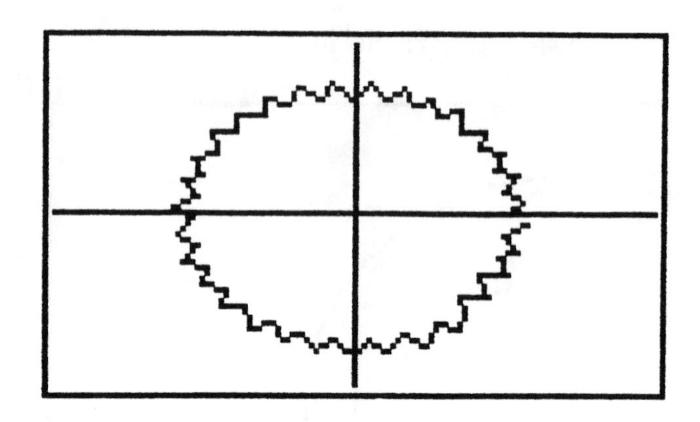

Aztec Design.

$$r1 = 3 + ((^{-}1) \wedge (\text{int}(16\theta/\pi)))$$

with $[0, 2\pi]_\theta$, θ-step .1, $[^{-}6, 6]_x$ and $[^{-}4, 4]_y$.

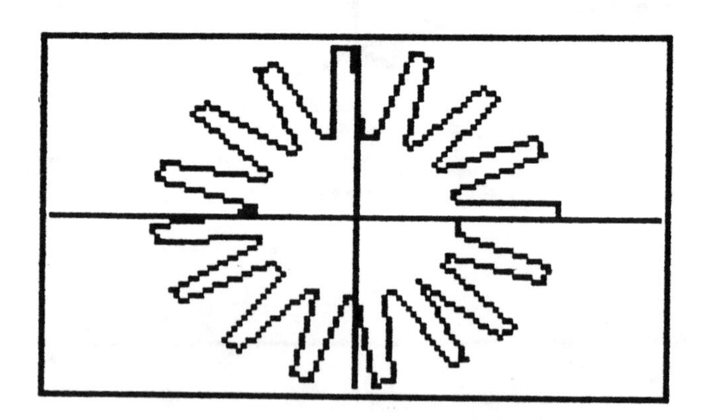

Pinwheels.

$$r1 = 4((30\theta/\pi) - \text{int}(30\theta/\pi))$$

with $[0, 2\pi]_\theta$, θ-step .1, $[^{-}6, 6]_x$ and $[^{-}4, 4]_y$.

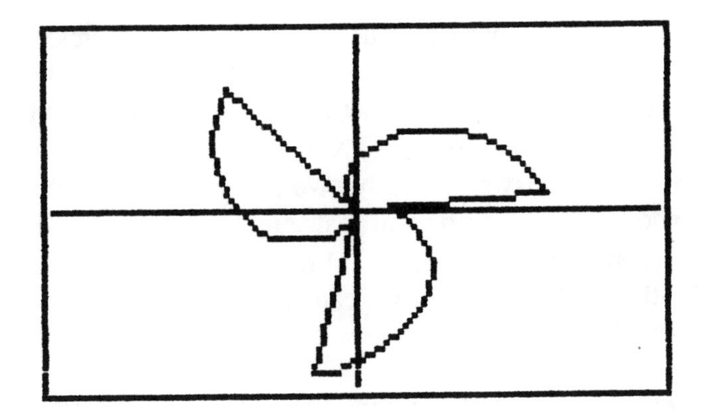

(The three preceding designs are courtesy of Professor Edward H. H. Gade III of the University of Wisconsin-Oshkosh Mathematics Department.)

Motion Problems

We conclude this section with two motion problems that can be modeled nicely on either the TI–81 or TI–85 using parametric graphing. The first uses horizontal-motion simulation; the second uses vertical-motion simulation with a variation.

Problem 1

Chris leaves Chicago on a fast train averaging 75 miles per hour bound for Los Angeles, about 2000 miles away. Kathy leaves Chicago by plane 20 hours later, traveling 400 miles per hour. Will Kathy arrive before or after Chris?

Calculator Solution. Set the **2nd MODE** to **Param**. Go to the **GRAPH** menu and set **GRAPH MORE FORMT** to **SimulG**. Now press **F1** (E(t) = menu) and define the following two sets of parametric equations:

xt1 = **75t**
yt1 = **4**

xt2 = **400(t − 20)**
yt2 = **2**

The x's are the position functions, and the y's represent horizontal paths. Now we set the **RANGE** settings at $[0, 33]_t$ with t-step .1, $[0,2000]_x$ with scale 0 and $[0, 6]_y$ with scale 0 for a pleasing picture of the paths. Now press **F5 (GRAPH)** and we obtain our answer visually. We can use **TRACE** and the up/down ▲▼ cursor buttons to determine the time and location at which Kathy passed Chris. Also, changing the t-step changes the time increments, so we can speed up or slow down the simulation.

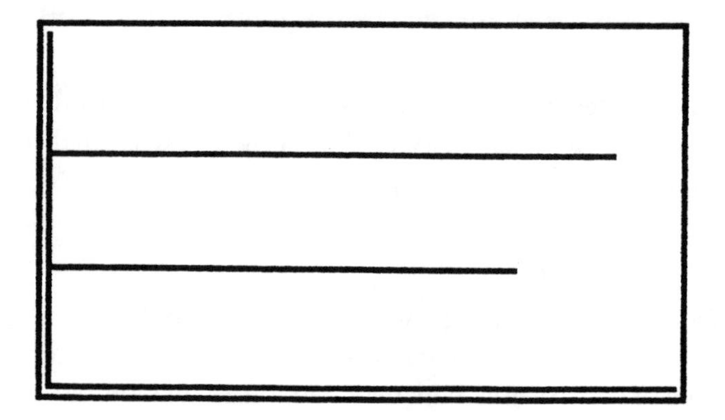

Problem 2

A model rocket is fired into the air from ground level. After .83 seconds its fuel is consumed; it is at a position 123 meters above the ground and it is traveling at a rate of 110.25 meters per second.

(a) When is the rocket 500 meters above the ground?

(b) What is the maximum height attained by the rocket?

(c) How much time elapses between the rocket's launch and its return to earth?

(d) After solving Parts (a), (b), and (c), change the problem to account for a horizontal wind of 6 miles per hour. Solve Parts (a), (b), and (c) under this new assumption.

Calculator Solution. We will produce the motion equation at the time ($t = 0$) when the rocket is out of fuel. Thus

$$s(t) = {}^{-}4.9t^2 + (110.25)t + 123$$

Suppose we represent the path vertically along the line $x = 2$. We need to have some sense of how many seconds it takes to reach maximum height and return to earth. A quick estimate would be about 25. (We can use the TI-85 **2nd POLY** menu to find roots for the position function.) So we define our parametric equations as:

$$xt1 = \mathbf{2}$$
$$yt1 = \mathbf{{}^{-}4.9t^2 + (110.25)t + 123}$$

and set our **RANGE** at $[0, 25]_t$ with t-step .1, $[0, 4]_x$ with scale 0, and $[0,800]_y$ with scale 0. [For a different effect, set t-step at .3 and GRAPH FORMT to **DrawDot**.]

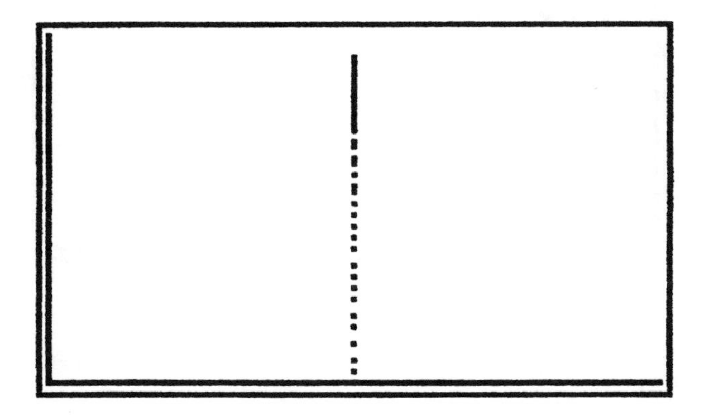

Now that we have a reasonable simulation of this vertical motion, we can answer Parts (a), (b), and (c) by **TRACE**ing along the graph and noting parameter values. (Use left ◀ or right ▶ cursor keys.)

(a) At about 4.2 seconds and again at about 18.3 seconds.

(b) About 743 meters (using t-step at .1).

(c) About 23.6 seconds.

Finally, to answer Part (d), we need to convert the xt1 equation to account for the wind. But the wind is given in miles per hour and our other velocity units are in meters per second. So we go to the Home screen, enter 6, then press **2nd CONV** (conversion menu), press **MORE** twice, select **SPEED**, select **mi/hr**, then select **m/s**, press **ENTER**, and we obtain 2.68224. That is, 6 mi/hr converts to 2.68224 meters/sec.

Now we return to **GRAPH**, select **E(t)** = and redefine xt1 = 2.68224t. Before we graph, however, we need to adjust the x-settings in the range (to accomodate for the horizontal drift of the rocket). Set x from 0 to 70 for a reasonable simulation, and then **GRAPH**.

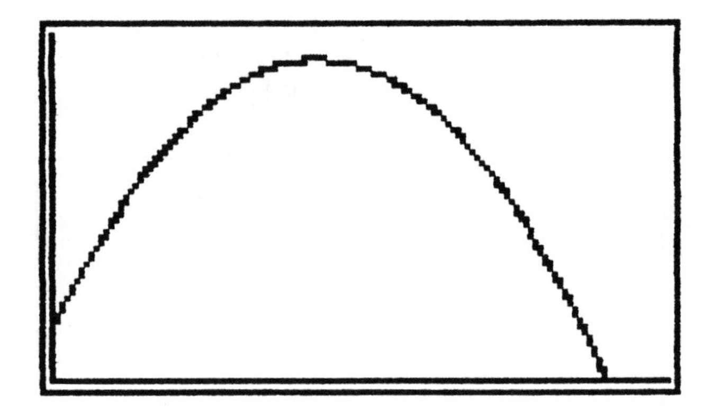

Formats of Common Functions/Operations with Multiple Arguments from Home Screen or in Programs

FORMAT	COMPUTES
arc (function, variable, xbeg, xend)	arc length of graph from beg x to end x
aug (matrix1, matrix2)	augments matrix1 with matrix2
Circl (x, y, r)	draws circle with center (x, y) radius r
cross (vector1, vector2)	cross product of vector1 with vector2
der1 (function, variable, value)	1st derivative of function at value of variable
der2 (function, variable, value)	2nd derivative of function at value of variable
dot (vector1, vector2)	dot product of vector1 with vector2
DS< (variable, real number)	programming: decrement variable by 1; skip next command if variable < number
evalF (function, variable, value)	evaluates function at value of variable

ExpR (x list, y list)	exponential regression on data from x, y lists [$y = ab^x$]
fMax (function, variable, lower, upper)	x-value of maximum of function between lower and upper bounds
fMin (function, variable, lower, upper)	x-value of minimum of function between lower and upper bounds
fnInt (function, variable, lower, upper)	definite integral of function from lower to upper limits of integration
For (variable, beg, end, increment)	programming: loop command
gcd (integer, integer)	greatest common divisor of two nonnegative integers
Hist (x list, frequency list)	draws histogram
InpSt (string, name)	programming: display string, ask for input, and store in name
inter (x_1, y_1, x_2, y_2, x)	interpolated (or extrapolated) y-value given 2 points (x_1, y_1), (x_2, y_2) and x-value
IS> (variable, real number)	programming: increment variable by 1; skip next command if variable > number
lcm (integer, integer)	least common multiple of two nonnegative integers
Line (x_1, y_1, x_2, y_2)	draws line from (x_1, y_1) to (x_2, y_2)
LinR (x list, y list)	linear regression on data from x, y lists [$y = a + bx$]
LnR (x list, y list)	logarithmic regression on data from x, y lists [$y = a + b \ln x$]
max (number, number)	larger of two numbers (can also use a list)
min (number, number)	smaller of two numbers (can also use a list)
mod (real number, real number)	modulo value of first real number with respect to second

mRAdd (number, matrix, integer1, integer2)	multiply row (integer1) of matrix by number, add to row (integer2) and store there
multR (number, matrix, integer)	multiply row (integer) of matrix by number, and store there
nDer (function, variable, value)	approximate numerical derivative of function at value of variable
OneVar (x list, frequency list)	one-variable statistical analysis on these lists
Outpt (integer1, integer2, message)	programming: display text message between text row (integer1) and text row (integer2)
pEval (list, value)	evaluate polynomial with coefficients in list (descending order) at value
P2Reg (x list, y list)	quadratic polynomial regression on data from x, y lists [$y = a_2x^2 + a_1x + a_0$]
P3Reg (x list, y list)	cubic polynomial regression on data from x, y, lists [$y = a_3x^3 + a_2x^2 + a_1x + a_0$]
P4Reg (x list, y list)	quartic polynomial regression on data from x, y lists [$y = a_4x^4 + a_3x^3 + a_2x^2 + a_1x + a_0$]
PtChg (x, y)	change point (on/off) status at (x, y)
PtOff (x, y)	erase point at (x, y)
PtOn (x, y)	draw point at (x, y)
PwrR (x list, y list)	power-model regression on data from x, y lists [$y = ax^b$]
rAdd (matrix, integer1, integer2)	add row (integer1) of matrix to row (integer2) and store there
randM (integer1, integer2)	display random matrix with dimensions integer1 × integer2

round (data structure, integer n)	rounds all entries of data structure to integer n ($0 \leq n \leq 11$) decimal places
rSwap (matrix, integer1, integer2)	interchanges two rows of matrix
Scatter (x list, y list)	draws scatter plot of statistical data from x, y lists
seq (term, variable, beg, end, increment)	list of sequence term values corresponding to variable from beginning value to ending value with increment
Shade (function1, function2, x_1, x_2)	shade area above function1, below function2, over interval $[x_1, x_2]$
simult (matrixA, matrixB)	solution vector X for system of simultaneous equations $AX = B$
Solver (equation, variable, number, {a, b})	solves equation for variable, with number as guess within bounds a to b
Sortx (x list, y list)	sort data in order of x elements
Sorty (x list, y list)	sort data in order of y elements
St ▸ Eq (string, equation)	convert string to equation and store as equation
sub (string, integer1, integer2)	subset of string from position integer1 to position integer2
TanLn (function, value)	draws tangent line to function at x = value
xyLine (x list, y list)	draws line plot of data from x, y lists

Menu and Submenu Displays

One of the strongest features of the TI–85 is that it is a "menu-driven" calculator. This, paradoxically, is also a source of confusion when one is first learning this calculator. There are so many menus, submenus, and options therein that it is frequently difficult to remember "where" a given operation, function, or command is. For example, the **GRAPH** key, alone, leads to 72 different options. The pages that follow contain displays of eleven of these internal menus. These pages, which illustrate various graphics features, can be copied directly to transparencies for overhead projection in class or in workshops. The following displays include:

• MODES	(2nd **MODE** key)
• VARIABLES	(2nd **VARS** key)
• GRAPHICAL	(**GRAPH** key)
• MATHEMATICAL	(2nd **MATH** key)
• MATRICES	(2nd **MATRX** key)
• STATISTICS	(**STAT** key)
• VECTORS	(2nd **VECTR** key)
• STRINGS AND LISTS	(2nd **STRNG** key) (2nd **LIST** key)
• PROGRAMMING	(**PRGM** key)
• PARAMETRIC GRAPHING	(2nd **MODE, Param, GRAPH**)
• POLAR GRAPHING	(2nd **MODE, Pol, GRAPH**)

TI-85

MODES

Normal	Sci		Eng
Float	0 1 2 3 4 5 6 7 8 9 0 1		
Radian		Degree	
RectC		PolarC	
Func	Pol	Param	DifEq
Dec	Bin	Oct	Hex
RectV	CylV	SphereV	
dxDer1	dxNDer		

TI-85

VARIABLES

2nd VARS MENU

ALL	REAL	CPLX	LIST	VECTR
MATRX	STRNG	EQU	CONS	PRGM
GDB	PIC	STAT	RANGE	

TI-85

GRAPHICAL

GRAPH MENUS, FIRST 5

y(x)=	RANGE	ZOOM	TRACE	GRAPH
define up to 99 functions	xMin	BOX		
	xMax	ZIN		
	xScl	ZOUT		
x	yMin	ZSTD		
y	yMax	ZPREV		
INSf	yScl	ZFIT		
DELf		ZSQR		
SELCT		ZTRIG		
ALL+		ZDECM		
ALL-		ZRCL		
		ZFACT		
		ZOOMX		
		ZOOMY		
		ZINT		
		ZSTO		

TI-85

GRAPHICAL

GRAPH MENUS, REMAINING 8

MATH	DRAW	FORMT		STGDB
LOWER	Shade	RectGC	PolarGC	RCGDB
UPPER	LINE	CoordOn	CoordOff	
ROOT	VERT	DrawLine	DrawDot	EVAL
dy/dx	CIRCL	SeqG	SimulG	
∫f(x)	DrawF	GridOff	GridOn	STPIC
FMIN	PEN	AxesOn	AxesOff	
FMAX	PTON	LabelOff	LabelOn	RCPIC
INFLC	PTOFF			
YICPT	PTCHG			
ISECT	CLDRW			
DIST	TanLn			
ARC	DrInv			
TANLN				

TI-85

MATHEMATICAL

2nd MATH MENUS

NUM	PROB	ANGLE	HYP	MISC	INTER
round	!	°	sinh	sum	INTERPOLATE
iPart	nPr	r	cosh	prod	x1 =
fPart	nCr	'	tanh	seq	y1 =
int	rand	▶ DMS	$sinh^{-1}$	lcm	x2 =
abs			$cosh^{-1}$	gcd	y2 =
sign			$tanh^{-1}$	▶ Frac	x =
min				%	y =
max				pEval	
mod				$\sqrt[x]{\ }$	SOLVE
				eval	

TI-85

MATRICES

2nd MATRX MENUS

NAMES	EDIT	MATH	OPS	CPLX
		det	dim	conj
		^{T}	Fill	real
		norm	ident	imag
		eigVl	ref	abs
		eigVc	rref	angle
		rnorm	aug	
		cnorm	rSwap	
		LU	rAdd	
		cond	multR	
			mRAdd	
			randM	

TI-85

STATISTICS

STAT

CALC	EDIT	DRAW	FCST	VARS
xlist Name=	xlist Name=	HIST	FORECAST:	$\bar{x}$
ylist Name=	ylist Name=	SCAT	x=	σx
▼	xStat	xyLINE	y=	Sx
x=Name y=Name	yStat	DRREG		$\bar{y}$
1-VAR	Names	CLDRW	SOLVE	σy
LINR		DrawF		Sy
LNR		STPIC		Σx
EXPR		RCPIC		Σx^2
PWRR				Σy
P2REG				Σy^2
P3REG				Σxy
P4REG				RegEq
STREG				corr
				a
				b
				n
				PRegC

TI-85

VECTORS

2nd VECTR MENUS

NAMES	EDIT	MATH	OPS	CPLX
		cross	dim	conj
		unitV	Fill	real
		norm	▶Pol	imag
		dot	▶Cyl	abs
			▶Sph	angle
			▶Rec	
			li▶vc	
			vc▶li	

TI-85

STRINGS AND LISTS

2nd STRNG

" sub lngth Eq ▸ St St ▸ Eq

2nd LIST

{ } NAMES EDIT OPS

OPS:

dimL
SortA
SortD
min
max
sum
prod
seq
li ▸ vc
vc ▸ li
Fill

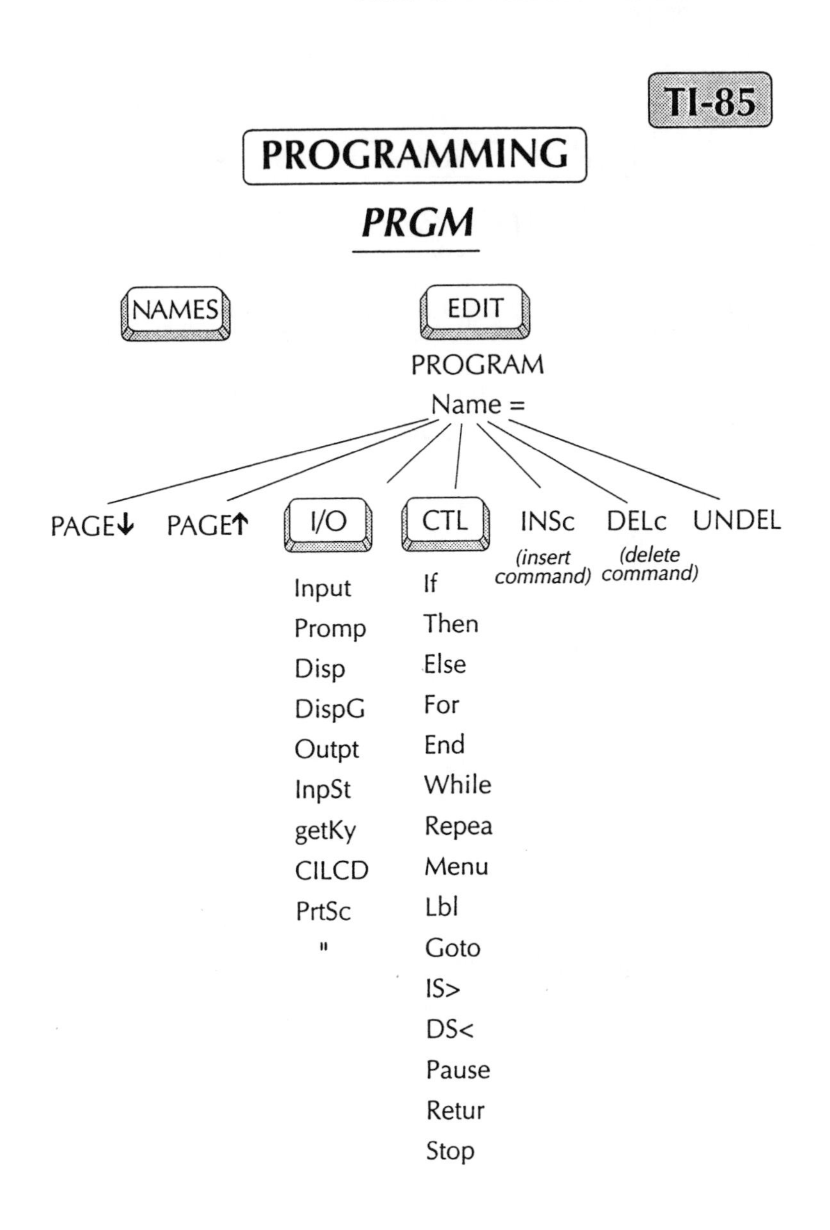
TI-85
PROGRAMMING
PRGM
NAMES
EDIT
PROGRAM
Name =
PAGE↓
PAGE↑
I/O
CTL
INSc
DELc
UNDEL
(insert command)
(delete command)
Input
Promp
Disp
DispG
Outpt
InpSt
getKy
CILCD
PrtSc
"
If
Then
Else
For
End
While
Repea
Menu
Lbl
Goto
IS>
DS<
Pause
Retur
Stop

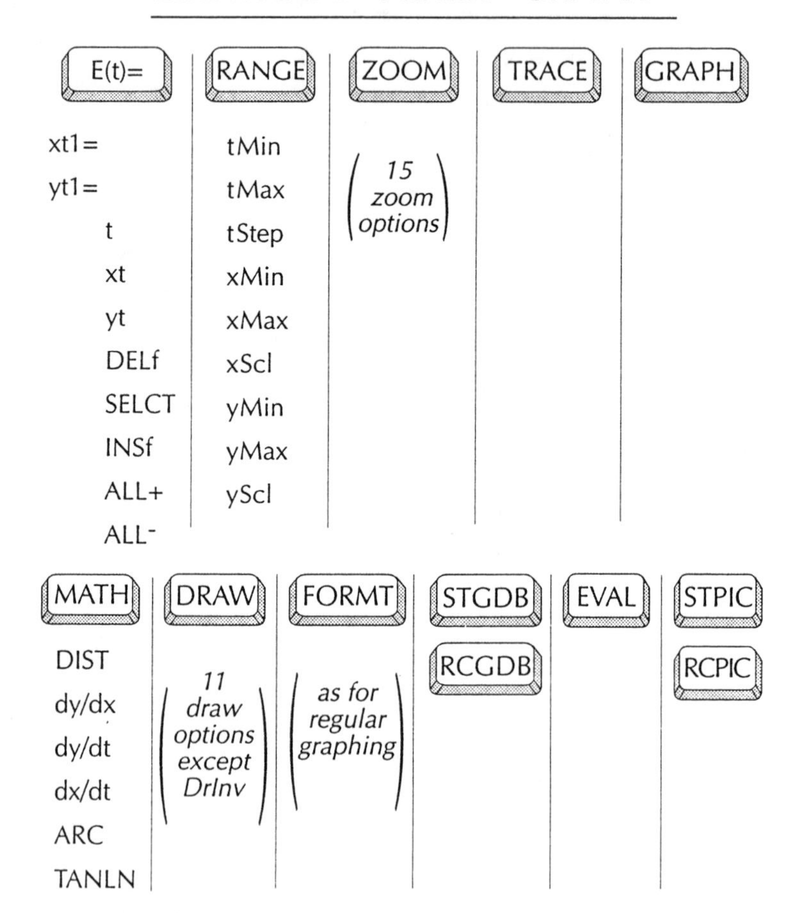
TI-85
PARAMETRIC GRAPHING
2nd MODE Param GRAPH
E(t)=
xt1=
yt1=
t
xt
yt
DELf
SELCT
INSf
ALL+
ALL-
RANGE
tMin
tMax
tStep
xMin
xMax
xScl
yMin
yMax
yScl
ZOOM
15 zoom options
TRACE
GRAPH
MATH
DIST
dy/dx
dy/dt
dx/dt
ARC
TANLN
DRAW
11 draw options except DrInv
FORMT
as for regular graphing
STGDB
RCGDB
EVAL
STPIC
RCPIC

TI-85

POLAR GRAPHING

2nd MODE Pol GRAPH

r(θ)=	RANGE	ZOOM	TRACE	GRAPH
θ	θMin	*15 zoom options*		
r	θMax			
INSf	θStep			
DELf	xMin			
SELCT	xMax			
ALL+	xScl			
ALL⁻	yMin			
	yMax			
	yScl			

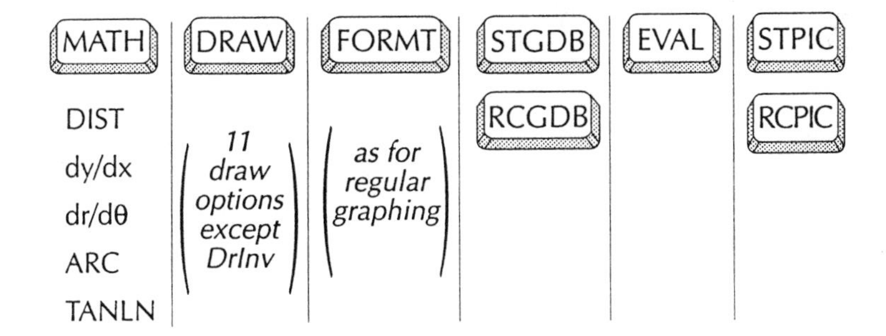